THE AI LEADER'S RENAISSANCE: DELEGATE THE ROUTINE, ELEVATE THE GENIUS

THE 5-STEP AI STRATEGY FOR BUSINESS LEADERS TO RECLAIM CONFIDENCE, PRODUCTIVITY, & COMPETITIVE EDGE IN 60 DAYS OR LESS

AUSTIN CHEN

BLACKSTONE PUBLICATIONS

Published by Blackstone Publications.

CONTENTS

INTRODUCTION

You notice it before your coffee is even finished.

A headline.

A post.

A colleague celebrating how they automated a task that used to take an entire afternoon.

AI doubled their output.

AI streamlined their workflow.

AI unlocked a new level of efficiency.

You keep scrolling, but the punch has already landed.

Because somewhere inside—quietly, honestly—you feel the sting of the question you don't want to ask:

"If machines can do the work... what exactly is my value now?"

You're not alone. Professionals everywhere are waking up with the uneasy sense that the ground is shifting beneath their feet faster than they can adjust.

Not because they're unwilling.

Not because they're incapable.

But because for the first time in their career, their expertise feels like it has an expiration date.

And the fear isn't loud.

It's subtle.

It shows up in small, private moments:

- The morning LinkedIn scroll that leaves you more anxious than informed.
- The team meeting where younger colleagues speak fluent AI while you're still translating the jargon.
- The bookmarked tutorials you never finish, the late-night experimenting that feels more like survival than growth.
- The internal whisper: *"I should be further along by now."*

Because beneath every post about "AI skills" is a deeper pressure:

You're not just learning a new tool.

You're trying to protect your professional identity.

That's the part no one talks about—the weight of watching the work you once took pride in suddenly feel... replaceable.

The dread that your hard-won abilities are depreciating faster than you can update them.

The fear that if you're no longer the one producing the work, you're no longer the one who matters.

But here's the truth you haven't heard enough:

Your problem isn't that you're behind.

Your problem is that you've been told to compete with the wrong thing.

Every "10X your productivity with AI" post pushes you toward a race you will never win.

Machines will always be faster, cheaper, and more efficient at mechanical output.

That is not your arena.

Your greatest value was never in the doing.

It has always been in the thinking—in your judgment, your interpretation, your discernment, your insight.

AI hasn't reduced your value.

It has revealed how much of your day is consumed by work unworthy of your potential.

Because the future does not belong to professionals who master every tool.

It belongs to the ones who learn how to **direct** these tools—freeing their time, expanding their creativity, strengthening their strategic judgment, and separating themselves from the noise.

This book gives you a clear, streamlined path to do exactly that.

Not another certification.

Not a catalog of tools that will be outdated by the time you learn them.

Not a reinvention that requires more hours than you have to give.

Instead, you'll adopt five strategic shifts that immediately change how you work:

- **Release the Routine** — offload repetitive, predictable tasks to AI and reclaim real hours of mental bandwidth.

- **Expand the Creative** — use AI to generate possibilities you shape, elevating your originality instead of threatening it.

- **Partner in Analysis** — process information at a scale no human could, while using your judgment to decide what truly matters.

- **Protect the Relational** — double down on the human skills machines cannot replicate: empathy, influence, leadership, trust.

- **Elevate the Strategic** — use AI as a thinking companion

> that sharpens decisions, reveals blind spots, and expands your professional vision.

When you integrate these shifts, everything changes.

You stop trying to outrun technology—and instead turn it into the engine that amplifies your career.

The professionals who rise in this era will not be the fastest or the most technical.

They will be the ones who understand what deserves their time, what does not, and how to use AI to expand—not erase—their relevance.

If you've been carrying the quiet worry that you're falling behind... you're standing at the edge of the solution.

Your next chapter isn't about becoming more like a machine.

It's about becoming unmistakably, strategically, powerfully human.

What comes next isn't about becoming more efficient.

It's about becoming impossible to ignore.

1

THE OBSOLESCENCE TRAP

"It is not the strongest of the species that survives, nor the most intelligent that survives. It is the one that is most adaptable to change." — Charles Darwin

THE EMAIL ARRIVES AT 7:14 AM on a Monday.

Your company's newsletter, usually forgettable, contains a single line that stops your morning routine cold: "We've implemented AI-powered content generation for the marketing department, reducing production time by 73%."

You read it twice. Then a third time.

Seventy-three percent.

You've spent fifteen years building expertise in exactly this area. You've developed an eye for what works, a feel for the brand voice, an instinct for what resonates with your audience. That expertise took thousands of hours to cultivate. Thousands of drafts. Thousands of small failures that eventually became pattern recognition you couldn't articulate if someone asked.

And now a software subscription can do it 73% faster.

The newsletter moves on to other topics—office updates, quarterly targets, the usual administrative noise. But you're not reading anymore. You're calculating. If content production is 73% faster, how long before someone asks why the team needs as many people? How long before your expertise—the expertise that defines your professional identity—becomes "overhead" on a spreadsheet?

This question haunts millions of professionals right now. According to Pew Research Center, 72% of workers report anxiety about AI replacing their jobs. Among knowledge workers specifically—consultants, marketers, analysts, strategists who've built careers on intellectual output—that number climbs to 76%. Three out of four professionals in your category wake up carrying the weight you just felt reading that email.

The anxiety isn't irrational. The anxiety is information.

Here's what the newsletter didn't tell you: the professionals who will thrive aren't the ones who learn to use AI the fastest. They're not the prompt engineers or the early adopters or the people who can name every model update. The professionals who will thrive are the ones who learn something harder: what to delegate and what to protect.

Technical mastery is the wrong goal. Strategic discernment is the only future-proof skill.

THE COMMODITIZATION CRISIS

The instinct, when faced with AI anxiety, is to compete. Learn the tools. Master the prompts. Become faster than the machine or at least fast enough to stay relevant.

This instinct will destroy your career.

Not because AI tools aren't worth learning—they are. But because competing with AI on speed, volume, or basic accuracy means competing on terms where you've already lost. If you define your professional value by how quickly you can produce a first draft, you are already obsolete. The machine does that faster. The machine will always do that faster. Next year's machine will be faster still.

Time-tracking studies consistently show professionals allocating

60-70% of their hours to activities that feel necessary but create no lasting value—coordination, status updates, formatting, search and retrieval. When researchers categorize how knowledge workers spend their days, they find a consistent pattern: most hours go to activities that are repetitive, predictable, and rules-based—exactly where machines excel.

Every task you can measure in units per hour is a task where you're competing with something that never sleeps, never gets tired, and improves continuously without asking for a raise.

Peter Drucker, the father of modern management, anticipated this moment. "The most important, and indeed the truly unique, contribution of management in the 20th century was the fifty-fold increase in the productivity of the manual worker," he wrote in 1999. "The most important contribution management needs to make in the 21st century is similarly to increase the productivity of knowledge work and knowledge workers."

Drucker's diagnosis was precise: knowledge work productivity had stagnated while manual work productivity soared. The assembly line transformed manual work by separating thinking from doing. The engineer designed the process; the worker executed the steps. This separation wasn't degrading—it was amplifying. It meant their physical capacity could produce far more value than ever before.

Drucker argued that knowledge work needed a similar transformation. But for decades, no tool could separate 'thinking' from 'doing' in cognitive tasks. The consultant had to write the slides. The analyst had to build the model. The marketer had to draft the copy. Thinking and doing remained fused.

AI is the assembly line for knowledge work. Not in the dystopian sense of replacing human minds—in the productive sense of allowing the separation that drives efficiency. The knowledge worker can now focus on the thinking: the strategy, the judgment, the direction. The doing—the drafting, the initial analysis, the routine synthesis—can be delegated.

Drucker's fifty-fold productivity increase becomes possible when knowledge workers stop doing the work themselves and start directing the work. That's not a future possibility. It's a current choice.

The trap deepens when professionals try to use AI to become "average faster." They plug ChatGPT into their workflow, generate outputs at higher volume, and feel productive. The emails go out quicker. The drafts accumulate. The meeting prep takes half the time.

But 'faster average' is still average.

When everyone has access to the same AI tools—and they do, or will soon—producing average work quickly just floods the market with average work. You're not differentiating yourself. You're not building anything defensible. You're racing to produce commodity outputs at commodity speeds, and there's always someone willing to do that cheaper.

This dynamic isn't new. It follows the pattern of every major technological shift in economic history.

During the Second Industrial Revolution, from roughly 1870 to 1914, craft production dropped from 80% of manufacturing output to 20%. Individual craftsmen who had defined themselves by their ability to make things—to execute, to produce, to create physical objects with their hands—found that capability devalued almost overnight. Machines could make things faster, cheaper, and more consistently.

Management roles increased eightfold during that same period. The economic value didn't disappear. It migrated. It moved from doing to directing, from execution to judgment, from manual production to strategic coordination.

The craftsman who insisted on competing with the factory—who tried to make things faster, to out-produce the machines—failed. The craftsman who recognized the shift and moved toward direction, coordination, and quality oversight—who became the person who decided what should be made and why—found an entirely new category of professional value.

You are living through the same transition, compressed into years instead of decades.

The signs of obsolescence are recognizable if you're honest with yourself. Look at your daily work and ask: Which of these activities could a competent professional handle with AI assistance in a fraction of the time?

Rote data entry. Basic email responses. Standard meeting notes. Initial research gathering. First-draft document creation. Routine scheduling. Status report generation. Information retrieval from existing documents.

If these activities consume a significant portion of your week, you're not building a defensible career. You're building a resume of tasks being automated faster than you can execute them.

THE PSYCHOLOGY OF DISPLACEMENT

The strategic case for delegation is straightforward. The emotional case is not.

Professionals don't just do tasks. They attach identity to tasks. The marketing director who writes campaign copy doesn't just produce words—she is a writer. The consultant who builds financial models doesn't just manipulate spreadsheets—he is an analyst. When AI takes over the production aspect of these activities, something deeper than efficiency is threatened.

This is Identity Erosion Anxiety: the fear that if you stop doing the thing, you stop being the thing.

A marketing director might rationally understand that AI can generate faster copy. But the question that surfaces at 3 AM isn't about speed or efficiency. It's existential: If I don't write the copy, am I still a strategist? If the machine does the analysis, am I still an analyst? What exactly do I contribute if the production part—the part I spent years mastering—is handled by software?

This fear deserves acknowledgment, not dismissal. Skill Relevance Terror is real. You spent years developing capabilities that now seem threatened. The intuition that took a decade to develop, the shortcuts that make you efficient, the pattern recognition that feels like instinct—all of it suddenly seems less valuable than a subscription anyone can purchase.

Seventy-six percent of knowledge workers carry some version of this anxiety. That number isn't irrational. It represents millions of people correctly identifying that the ground beneath their careers is shifting.

The fear becomes paralyzing when combined with the sheer volume of change. Over 14,000 new AI tools launched in 2023. That's 38 new AI tools per day. Every morning brings new capabilities, new platforms, new headlines about what AI can now accomplish that it couldn't accomplish yesterday.

'Tool Fatigue' is the natural response. When you can't possibly evaluate every new platform, when the landscape changes faster than you can map it, the brain does what brains do under conditions of overwhelm: it freezes. Analysis paralysis sets in. You read articles about AI productivity but don't actually use any tools. You attend webinars but don't implement the techniques. You feel increasingly behind while doing increasingly little to catch up.

The paralysis feels like protection but functions as acceleration. Every day you spend frozen is a day your competitors spend experimenting. Every month you spend overwhelmed is a month someone else spends building fluency with tools that will reshape your industry.

Breaking the paralysis requires a shift in framing. Not from ignorance to expertise—that path is impossible given the rate of change. But from victimhood to ownership.

"AI is happening to me" is the victim frame. It positions technology as something external, threatening, beyond your control. You're a passenger in someone else's transformation, watching helplessly as the landscape reorganizes itself around you.

"I am hiring AI as my staff" is the ownership frame. It positions technology as something you direct, deploy, and manage. You're not learning to use a tool. You're learning to delegate. You're not becoming more technical. You're becoming a more effective manager of technical resources.

The second frame doesn't require you to understand how large language models work. It doesn't require you to follow every product launch or evaluate every new platform. It requires you to understand what you need done—and to develop the judgment to know what should remain in human hands.

THE STRATEGY OF DISCERNMENT

Strategic discernment begins with a single sorting mechanism. Every task in your professional life belongs in one of two categories: work that should be automated and work that should be amplified.

But before you start sorting, a caution: the most common AI adoption mistake isn't resistance—it's premature automation.

Professionals hear about AI capabilities, identify tasks that sound automatable, and immediately start delegating. They skip the critical step: understanding why those tasks exist and what quality looks like in them.

When you automate a task you don't deeply understand, you can't evaluate whether the AI is doing it well. You can't craft effective prompts because you don't know what good outcomes require. You can't catch errors because you don't know what errors look like.

The result: AI produces output that seems fine but isn't. Emails go out with the wrong tone. Analysis misses key variables. Summaries emphasize the wrong points. The professional doesn't notice because they never really knew what they were looking for.

The correct sequence is always: master first, then delegate. You need to have done the task well enough, long enough, to recognize quality instantly. Then—and only then—can you direct AI effectively and review its output efficiently.

This doesn't mean spending years on every task before automating. It means being honest about where your expertise is deep enough to evaluate and where you're guessing. Automate from knowledge, not from hope.

With that foundation established, the sorting can begin.

The Automate category contains tasks that are repetitive, predictable, low-stakes, and rules-based. These are Robot Zone activities—not because they're unimportant, but because they don't require what makes you distinctively valuable. They need to be done. They don't need you specifically to do them.

Email responses that follow established patterns. Meeting summaries that extract key points from transcripts. Research that gathers information from known sources. Document formatting that applies consistent standards. Calendar coordination that follows clear preferences. Status updates that synthesize predictable data.

These activities share a common characteristic: a competent assistant with clear instructions could handle them. AI is that assistant —faster, cheaper, and available at 3 AM when you're trying to clear your inbox before tomorrow's meeting.

The Amplify category contains tasks that require empathy, complex context, high-stakes judgment, and human trust. These are Genius Zone activities—not because they're harder, but because they depend on capabilities that remain distinctively human.

Strategic decisions that require weighing incommensurable values. Negotiations that depend on reading emotional undertones. Creative work that expresses a genuine point of view. Relationship building that establishes authentic trust. Conflict resolution that requires navigating interpersonal dynamics. Judgment calls where the right answer depends on context no system can fully capture.

These activities share a different characteristic: they're not just intellectual exercises. They depend on your specific perspective, your accumulated experience, your capacity to care about outcomes in ways that matter to other humans.

Between these two zones lies a third: the Partner zone. This is the gray area where humans and machines work together in iterative collaboration.

Analysis that requires AI to process large datasets but needs human interpretation to extract meaning. Creative projects where AI generates options but humans provide direction and curation. Research where AI accelerates information gathering but judgment determines what information matters.

The Partner zone is where most interesting professional work will live. Not purely automated. Not purely human. But collaborative in ways that amplify human capability rather than replacing human involvement.

The skill you need—the only future-proof skill—is the judgment to sort activities correctly. The professional who delegates Robot Zone work to AI and reinvests that time in Genius Zone work builds an increasingly defensible position. The professional who clings to Robot Zone work out of habit or fear finds themselves competing on terrain where they cannot win.

This isn't about learning every AI tool. It's about developing the discernment to know which tools matter for your work, what those tools can actually accomplish, and where human judgment remains irreplaceable.

Technical mastery—understanding how transformers work, keeping up with every model release, becoming a prompt engineering expert—is optional. Strategic discernment is not.

YOUR COGNITIVE ENERGY AUDIT

Discernment requires data. Before you can delegate strategically, you need to see clearly where your time and energy actually go.

Every time you shift attention from one task to another, you lose twenty-three minutes. This number, from a University of California study, measures how long it takes to fully re-engage with the original task after an interruption. Not just to glance at it again—to reach the same depth of cognitive engagement you had before the switch.

Calculate what this means for a typical workday. Five context switches—email to document to meeting to email to analysis—costs nearly two hours to the switching itself. Those hours don't appear on any timesheet. They disappear into the friction between activities.

Now consider what AI delegation changes. When routine tasks don't require your full attention—because AI handles the first pass and you only review—you switch less. When you can batch the review of multiple AI outputs rather than creating each piece individually, you maintain focus longer.

The cognitive energy audit isn't just about identifying Robot Zone tasks. It's about identifying where context-switching is silently consuming hours that could go to sustained, deep work.

Twenty-three minutes per switch. Five switches per day. Ten hours per week. That's a quarter of your working time—recapturable.

Most professionals, when asked about their workweek, describe it in terms of roles and responsibilities. "I manage the marketing team." "I advise clients on digital strategy." "I oversee product development." These descriptions capture titles and job functions. They don't capture how hours actually disappear.

The Cognitive Energy Audit changes that. For three consecutive workdays, you'll track every task you perform—not in project management categories, but in energy categories. The goal isn't perfect accounting. The goal is honest observation.

Every thirty minutes, note what you're doing. Then answer three questions about that activity:

First, how repetitive is this task? Is this something you do daily, weekly, monthly, or rarely? High-repetition tasks are strong candidates for automation. If you're doing the same thing every day, the pattern is clear enough to delegate.

Second, what cognitive load does this require? Rate it from one to ten. A routine email response might be a 'two'. A complex strategic recommendation might be a nine. Low cognitive load tasks often masquerade as important work while draining energy that could go elsewhere.

Third, what strategic value does this create? Again, one to ten. Does this activity move your career forward? Does it strengthen relationships that matter? Does it produce outcomes that differentiate you from competitors? Or is it maintenance—necessary but not building anything?

Most professionals who complete this audit discover something uncomfortable: 40% to 60% of their working hours go to tasks that score low on strategic value while scoring high on repetition. They're spending their best energy on work that any competent professional could handle—and increasingly, that AI can handle better.

Consider the math. Fifty hours per week with forty percent going to

Robot Zone activities means twenty hours weekly spent on work that doesn't require you. Twenty hours that could go toward the strategic thinking, relationship building, and creative work that actually builds your career.

What would your professional life look like with twenty additional hours for Genius Zone work?

The strategy sessions you've been postponing because you're always reacting to immediate demands. The relationships you've been meaning to deepen but never have time to nurture. The creative projects that keep getting pushed to "someday" while routine tasks consume today. The learning you want to pursue but can't fit into an already-overloaded schedule.

The Cognitive Energy Audit doesn't tell you what to do. It shows you what you're doing—and that visibility changes everything.

When you see, quantified in actual hours, how much time goes to activities that don't require your specific expertise, the case for delegation stops being theoretical. You're not imagining a vague future where AI helps somehow. You're looking at fifteen or twenty hours per week that could be reclaimed through strategic automation.

That's the foundation for everything that follows. Before you can release routine work, you have to identify it. Before you can amplify your highest-value contributions, you have to know what those contributions are and how little time currently goes to them.

Complete the audit before moving forward. Track your hours. Answer the questions honestly. Let the data show you what intuition has been obscuring.

The numbers will tell you where to start.

But there's an objection you're likely forming right now. It sounds reasonable, even responsible: 'I don't have time to learn new systems'.

The irony is sharp enough to draw blood: you're too busy doing work to learn tools that would make you less busy.

You've seen the AI demos. You know the potential. But between the meetings and the emails and the deadlines, there's no space for experimentation. Learning new systems means slowing down today for a theoretical payoff tomorrow. And today has enough problems.

This objection is completely rational. It's also exactly why you're trapped.

The time debt compounds. Every week you spend doing work manually that could be delegated is a week you're investing in a skill set that's depreciating. It's like refusing to learn to drive because walking to work takes up all your time.

Here's what the belief of " having no time" misses: you don't need to learn AI. You need to use AI on one task.

The learning curve for strategic delegation is measured in minutes, not months. You don't need to understand transformer architecture or prompt engineering theory. You need to type a request, evaluate the output, and iterate if necessary. If you can write an email to a colleague, you can write a prompt to an AI.

The time objection evaporates when you stop thinking about "learning AI" and start thinking about "using AI once." The learning happens automatically through use. The time investment is negative—you get more time than you spend.

YOUR FIRST 72 HOURS

The full Cognitive Energy Audit takes three days of careful tracking. The complete framework covers four distinct task categories. The psychological resistance might take weeks to fully process.

But you don't need to do everything to start. You need to do one thing—and then another.

Here's the minimum viable path from awareness to action: seventy-two hours from reading this chapter to your first strategic delegation.

Day One: The Ten-Task Test

Before your workday begins, write down the ten tasks you expect to handle today. Don't filter—include everything from major projects to minor emails. Just list them.

At the end of the day, review your list. For each task, answer one question: "Could a competent new hire do this with clear instructions?"

If yes, the task is probably the Robot Zone. It requires execution, not judgment. A human with basic professional competence—or an AI with good prompting—could handle it.

If no, ask the follow-up: "Why not?" The answer reveals what makes certain work distinctively yours. Maybe it required relationship context. Maybe it needed creative judgment. Maybe the stakes were high enough that your specific expertise was essential.

Most professionals who complete this exercise find that six or seven of their last ten tasks could have been handled by that hypothetical new hire. They're spending the majority of their time on work that doesn't require their accumulated expertise—and wondering why they never have time for the work that does.

This isn't the full audit. But it's enough to see whether the pattern exists. If more than half your recent work falls in the "new hire could do it" category, the detailed audit will show you exactly where those hours are going.

Day Two: The Single Experiment

Choose one task from yesterday's "yes" list. Ideally, it should be:

Low stakes (internal, not client-facing) Repetitive (something you do regularly, not once) Time-bounded (completable in under an hour)

Before you do this task manually, try delegating it to AI.

Open whatever AI tool you have access to—ChatGPT, Claude, Copilot, any large language model. Describe what you need. Be specific about format, length, and purpose. Submit and wait.

When the output arrives, evaluate it. Not against perfection—against acceptability. Would this work? What would need to change?

If the output is usable with minor edits, you've just experienced strategic delegation. The time you saved is real. The quality delta is probably smaller than you expected.

If the output is unusable, you've learned something: this particular task may require more context than a simple prompt provides, or it may belong in a different category than you assumed. Both insights are valuable.

Either way, you've completed an experiment. The barrier between "understanding AI delegation" and "doing AI delegation" is now behind you.

Day Three: The Time Capture

For every task today, track two numbers: how long the task took with your current approach, and how long you estimate it would take with AI assistance (based on your Day Two experience or reasonable projection).

Don't change anything yet. Just track.

At the end of Day Three, add up the potential time savings. If your estimates suggest that AI delegation could reclaim two hours daily—not unusual for professionals with high routine task loads—you're looking at ten hours weekly. Forty hours monthly. Nearly five hundred hours annually.

What would you do with five hundred hours?

That question—answered specifically, not abstractly—is your motivation for continuing past these first seventy-two hours. Keep that answer visible. When delegation feels uncomfortable, when the old habits pull you back, the answer reminds you why the discomfort is worth it.

What You've Accomplished

After seventy-two hours, you have:

Visibility into your Robot Zone work (the tasks a new hire could handle) Personal experience with AI delegation (at least one completed experiment) Quantified potential (estimated time savings grounded in your actual workload) Motivation (a specific answer to "what would I do with the time?")

You haven't mastered strategic delegation. You've started. The full framework, the psychological work, the ongoing practice—that comes in the weeks ahead. But you're no longer theorizing about a future change. You're building on an actual beginning.

Seventy-two hours from now, you'll know whether this path is right for you. Not from thinking about it—from doing it.

You've diagnosed the problem. You've identified the trap. You understand that strategic discernment—not technical mastery—is the skill that matters. And now you have a concrete path to your first experiment with delegation.

But understanding is not the same as action, and action requires more than a 72-hour protocol.

The next chapter addresses the obstacle that stops most professionals before they begin: the psychological resistance to delegation itself. The guilt that says using AI is "cheating." The imposter syndrome that questions whether you're qualified to direct rather than do. The ethical concerns that need frameworks before you can proceed with confidence.

You know what needs to change. Now you need to change how you think about change itself.

2

WHEN EFFICIENCY FEELS WRONG

Professional culture operates on a fundamental equation: effort equals value. The harder you work, the more you're worth. The longer the hours, the greater the dedication. Struggle signals commitment. Difficulty demonstrates worth. We learn this early, internalize it deeply, and build entire careers on its foundation. Then AI appears and shatters the equation completely. Tasks that took three hours now take twenty minutes. Work that required visible struggle now happens with minimal friction. The outcome remains identical. The value to clients unchanged. But the sweat equity component—the struggle that proved you cared—vanishes. And with it, a familiar way of demonstrating professional legitimacy. This chapter addresses what happens when the equation breaks and how to build professional identity on something more sustainable than suffering.

THE IMPOSTER SYNDROME OF THE NEW ERA

Every professional transition creates the same anxiety. Accountants worried spreadsheets would make them obsolete. Writers feared word processors would destroy their craft. Managers thought email would cheapen communication. The pattern repeats: when difficulty disappears, people question their legitimacy.

We're in another cycle. AI makes hard work easier, and that ease triggers discomfort about professional worth. History shows what happens next. The professionals who clung to outdated difficulty disappeared. The ones who redirected their expertise toward higher-value work thrived.

The resistance shows up as a voice. Former professor. Parents. Senior colleague whose respect you've spent years earning. The voice asks: *If I use AI to write this, am I really the one writing it? Isn't this just cheating? What happens when someone finds out I'm not doing everything myself?*

This is "cheating" guilt. The sense that using AI diminishes your contribution. That efficiency is laziness in disguise. That real professionals do hard work manually because that's what makes it valuable.

The guilt has a source.

Professional culture spent decades reinforcing a specific equation: effort equals value. The student who stays up all night gets praised for dedication, even when the well-rested student produces better work. The employee who arrives first and leaves last gets recognized, regardless of output. The internalization runs deep: struggle signals commitment, difficulty demonstrates worth.

AI breaks this equation. When a task that took three hours takes twenty minutes, the effort-value connection snaps. The work still needs to meet standards. The outcome still matters to clients and stakeholders. But the sweat equity component—the visible struggle that proved you cared—evaporates.

This creates cognitive dissonance. You know intellectually that outcomes matter more than effort. You know clients pay for results, not for watching you suffer. But the emotional programming persists. Using AI feels like skipping steps. Like getting away with something.

Here's the reframe: efficiency isn't laziness. Efficiency is the prerequisite for high-value contribution.

When you spend three hours formatting a document that AI handles in ten minutes, you're not demonstrating dedication. You're demonstrating poor resource allocation. You're choosing to invest finite, irreplaceable cognitive energy in work that doesn't require it. Time spent on Robot Zone tasks is time stolen from Genius Zone contributions. From strategic thinking. From relationship building. From work that actually justifies your compensation and advances your career.

Using AI strategically isn't cheating. It's recognizing that your professional value lies in judgment, not manual execution. The architect doesn't pour concrete. The surgeon doesn't sterilize instruments. The director doesn't operate the camera. These professionals delegate execution because their expertise lies elsewhere—in design, in judgment, in vision. Their value increases when they focus on what only they can provide.

The same logic applies to you.

THE EXPERT FALLACY

The guilt travels with a companion: the belief that you need technical expertise before you can legitimately use AI. That you should understand transformers and attention mechanisms and temperature parameters before typing your first prompt. That "real" AI users are prompt engineers with computer science backgrounds.

This belief is false. It paralyzes people who should be benefiting from these tools right now.

You don't need to understand internal combustion to drive a car. You don't need to know TCP/IP protocols to send email. You don't need to understand large language model architecture to use AI effectively in your professional practice.

What you need is subject matter expertise. You need to know what "good" looks like in your field. You need judgment to evaluate whether AI output meets your standards. You need professional knowledge to ask the right questions and recognize when answers fall short.

This is expertise you already have. Fifteen years of experience didn't teach you how neural networks function. But it taught you exactly what a quality proposal contains, what tone works for difficult client conversations, what gaps exist in competitive analysis. That knowledge is the foundation of effective AI collaboration. The technical details are irrelevant to your success.

REDEFINING EXPERTISE

The shift required isn't from non-expert to expert. It's from creation-centered expertise to curation-centered expertise.

Traditional professional identity centers on creation from scratch. The strategist who writes the deck. The analyst who builds the model. The marketer who crafts the copy. Value is demonstrated through production—through visible, attributable output that bears your personal mark.

The new expertise centers on curation, direction, and verification. The strategist who shapes the narrative arc, regardless of who produces the first draft. The analyst who identifies the right questions, then validates whether the answers make sense. The marketer who defines the voice and message, then ensures output maintains brand integrity.

An architect doesn't lose professional standing because they didn't personally install the plumbing. They gain standing by ensuring the plumbing serves the building's purpose. Their expertise lies in vision, integration, quality standards—not in manual labor of pipe fitting.

Your expertise works the same way. When you direct AI to produce a draft, then reshape that draft through professional judgment, the resulting work is yours. Not because you typed every word. Because you applied the knowledge, taste, and strategic thinking that made those words effective.

BUILDING EARLY CONFIDENCE

Psychological resistance weakens through experience, not argument. Reading about AI capabilities doesn't build confidence. Using AI successfully does.

Start with private experiments where stakes approach zero. Personal tasks. Internal documents no one else will see. Low-priority items where imperfection carries no professional consequences. The goal isn't producing polished output. The goal is proving to yourself that you can control the tool—that it responds to your direction, that you can evaluate its output, that you can improve results through iteration.

Draft an email you'd normally labor over, then compare the AI version to what you would have written. Summarize a document you've already read to test whether the AI captures what matters. Generate five approaches to a problem you've already solved, then evaluate which suggestions have merit.

These experiments break abstract fear into concrete, manageable interactions. You discover the tool isn't magic—it's pattern recognition that responds predictably to clear instructions. You discover you have opinions about what it produces—preferences, criticisms, improvements. You discover your judgment still matters.

That discovery is the foundation for everything that follows.

THE FORGETTING FEAR

Behind the skill-loss concern sits something deeper: the worry that AI will erode your cognitive capability itself. Not just that you'll stop practicing certain tasks, but that you'll lose the ability to think without machine assistance.

This fear treats the mind like a muscle that atrophies uniformly from any reduction in use. Stop doing mental math, lose the ability to reason with numbers entirely. Stop writing first drafts, lose the capacity for original thought. AI dependence becomes cognitive disability.

The research doesn't support this catastrophic view. But it doesn't dismiss the concern entirely either.

What we know: cognitive abilities are specific, not general. Practicing mental arithmetic improves mental arithmetic. Limited transfer to other numerical reasoning. Writing first drafts improves first-draft writing. Doesn't automatically strengthen strategic thinking or critical analysis. The skills you stop practicing decline. Other cognitive capacities remain intact or improve when freed from maintenance tasks.

What we also know: higher-order thinking requires raw material.

You can't synthesize insights from information you never encountered. You can't evaluate quality without examples of quality to compare. If AI handling means you never engage deeply with content at all—if you outsource not just production but attention—then genuine cognitive concern is warranted.

The solution is engagement, not avoidance. When AI produces a draft, you don't skim and approve. You read critically, noting what works and what doesn't. When AI summarizes research, you spot-check sources and consider what might be missing. When AI suggests options, you evaluate each against your judgment rather than selecting the first acceptable choice.

AI-assisted work can actually increase cognitive engagement compared to pure manual production. When you're not exhausted from the labor of first-draft creation, you have energy for higher-order thinking that produces quality. When you're not drowning in routine tasks, you have bandwidth for strategic reflection that produces insight.

The "forgetting how to think" fear is real in one scenario: passive consumption of AI output without genuine mental engagement. The antidote isn't avoiding AI. It's ensuring your use of AI involves more thinking, not less. Delegation of execution should increase cognitive investment in evaluation, direction, and judgment. If it doesn't—if AI is just making you lazy—the problem isn't the tool. It's the approach.

ETHICAL FRAMEWORKS TO CONSIDER

Psychological resistance isn't the only barrier. Legitimate ethical concerns also require resolution before you can integrate AI confidently into professional practice.

The concerns are real. Questions about transparency, data privacy, accuracy, and authenticity deserve serious answers—not dismissal as obstacles to overcome. Building an explicit ethical framework transforms these concerns from paralyzing uncertainties into clear decision criteria.

The question arises immediately: when do you disclose AI assistance?

There's no universal answer. But there are useful distinctions.

Backend thinking rarely requires disclosure. Brainstorming sessions, research organization, draft generation for internal use, personal productivity enhancement—these processes don't typically demand transparency about which tools you used. When you use a calculator for financial modeling, you don't disclose "calculator-assisted analysis." The principle extends to AI assistance in preparation and thinking work.

Final deliverables occupy different territory. When output carries your professional name and goes to clients, stakeholders, or the public, disclosure questions become more nuanced. The answer depends on industry norms, client expectations, and the nature of the work.

Some contexts demand transparency. Academic submissions, legal filings, regulated industries often have explicit rules about AI disclosure. Follow them.

Other contexts depend on expectations. If a client is paying specifically for your personal analysis—your brain, your perspective, your years of experience applied to their problem—heavy AI reliance might misrepresent what they're receiving. If they're paying for a deliverable that meets certain standards, the production method matters less than quality.

The practical approach: when uncertain, err toward disclosure. A simple statement—"I used AI assistance in drafting this document, with full human review and editing"—resolves ambiguity without diminishing your contribution. Most clients care far more about quality than production methods. Many appreciate transparency.

Here's a technique that eliminates disclosure anxiety entirely: default to transparency in low-stakes contexts. When you share a document internally, mention casually: "I used Claude to generate the first draft, then revised heavily." When you post on LinkedIn, add a note: "AI-assisted writing, human-directed thinking." When you send an email summary, note in the signature: "Summary generated with AI assistance."

Do this consistently for three weeks. Two things happen.

First, you discover almost no one cares. The reactions you feared—

judgment, suspicion, diminished respect—don't materialize. What materializes is curiosity. People want to know what tools you're using. They want tips on how to do the same. Your transparency positions you as a resource, not a suspect.

Second, you internalize that AI use isn't shameful. The act of saying it out loud—repeatedly, routinely—rewires the internal narrative. What felt like confession becomes information. What felt risky becomes normal. By the time a genuinely high-stakes disclosure decision arises, you've already built the muscle memory for matter-of-fact transparency.

Data Privacy

Some rules are non-negotiable.

Never enter personally identifiable information about real individuals into public AI systems. Client names, employee details, health information, financial data tied to specific people—none of it belongs in a tool you don't control.

Never enter trade secrets, proprietary strategies, or confidential business information. Your company's competitive advantages, unpublished financial data, internal strategic plans—assume anything you type could become training data for a system that serves your competitors.

Never enter content you're contractually obligated to protect. NDA-covered information, attorney-client communications, classified material—the convenience of AI assistance doesn't override your professional and legal obligations.

This matters more than professionals realize. According to enterprise surveys, 11% of employees have entered confidential company data into ChatGPT. 5% have entered client data. These aren't hypothetical risks. They're documented patterns of behavior that create real liability.

The safe approach: treat AI tools like a brilliant but untrustworthy intern with an internet connection. Useful for general tasks, never to be trusted with information you wouldn't want published.

Anonymize examples. Generalize details. Protect what needs protecting.

The Human-in-the-Loop Standard

One rule supersedes all others: no AI-generated content leaves your desk without human review.

This is the difference between a professional and a spammer. Spammers use AI to generate volume without accountability. Professionals use AI to generate drafts that undergo genuine evaluation before reaching anyone else.

Human review means actually reading the output. Not skimming. Not assuming. Not hitting send because you're busy. It means checking facts that can be checked. It means evaluating whether the tone matches your intention. It means asking whether you'd be comfortable defending this content if questioned about specific claims or phrasing.

The standard isn't perfection. It's accountability. You remain responsible for everything that carries your name—which means everything deserves the attention that responsibility requires.

Recognizing Bias and Blandness

AI outputs share a consistent characteristic: smooth competence masking significant limitations.

The text reads well. Sentences flow. Grammar is correct. The surface appearance of quality is nearly automatic. This surface quality is precisely the danger—because it makes flawed content easy to accept without examination.

AI defaults toward the average. It's trained on vast amounts of text, which means it reproduces patterns that appear frequently across that training data. Common phrasings. Conventional structures. Middle-of-the-road positions on contested topics. The output sounds professional precisely because it sounds like everything else.

Your job is to catch what the average misses.

Are the facts accurate, or plausible-sounding fabrications? AI generates confident assertions whether or not those assertions correspond to reality. Dates, statistics, citations, names—all require verification against actual sources.

Does the analysis have a genuine point of view, or is it hedged into meaninglessness? AI often produces content that acknowledges all sides without committing to a position. Useful for some purposes. Useless when you need actual recommendations.

Is the voice distinctive, or could anyone have written this? AI-generated content tends toward a particular register—smooth, professional, forgettable. Your expertise includes a perspective, a style, an approach that differentiates you. The AI draft is raw material. Your revision adds what makes it yours.

Your critical eye is the safety mechanism. Without it, you're not collaborating with AI—you're laundering generic content through your professional reputation.

With psychological barriers addressed and ethical frameworks established, one obstacle remains: the gap between understanding and action. You know you should experiment with AI. You know it could help. You've resolved the guilt and clarified the ethics.

But you haven't actually started.

The laboratory mindset bridges this gap. It transforms AI adoption from a high-stakes commitment into a series of low-consequence experiments. It replaces the pressure to "get it right" with the freedom to discover what works.

Low-Stakes Sandboxes Begin where failure costs nothing.

Personal tasks make ideal starting points. Draft a thank-you note. Summarize an article you want to remember. Generate gift ideas for a difficult-to-shop-for relative. Plan a vacation itinerary. Organize notes from a book you're reading.

These experiments teach you how AI responds without professional risk. You discover its tendencies—where it excels, where it struggles, what instructions produce useful output. You develop intuition about prompting that transfers to professional applications. You build confidence through accumulated small successes.

Internal documents come next. Meeting agendas. Status reports. Project documentation. Work that matters enough to be useful practice, but remains within your organization and under your control. If the output needs significant revision, no client relationship suffers. If the experiment fails entirely, you've lost only the time invested.

Client-facing work comes last, and only after you've developed reliable judgment about AI output in your domain. The progression isn't arbitrary caution—it's strategic skill building. You wouldn't present to the board without rehearsal. You shouldn't present AI-assisted work to clients without developing the evaluation skills that ensure quality.

The Draft Zero Approach

One of AI's highest-value applications requires no technical sophistication: breaking the terror of the blank page.

Every professional knows the experience. A document needs writing. You know what it should contain, roughly. But the cursor blinks on an empty screen, and starting feels impossible. You check email. Refill coffee. Find urgent tasks that suddenly need attention. The document waits.

AI eliminates this resistance entirely.

You don't need the AI to produce final copy. You need it to produce something—anything—that you can react to. Draft Zero isn't meant to be good. It's meant to be a starting point that your expertise can improve.

The instructions can be simple: "Write a first draft of a project proposal covering timeline, budget, team structure, and key milestones." The output will be generic. It will miss important nuances. It will probably include things you'd never actually say.

That's fine. Because now you're editing instead of creating. Editing is psychologically easier than creating. The blank page is gone. You're responding to something concrete, exercising judgment about what works and what doesn't, making specific improvements to specific sentences.

The final document might retain 20% of the Draft Zero language. Or 80%. Or nothing at all—you might delete everything and start fresh, but now you're writing with momentum because you've already engaged with the problem. The AI draft served its purpose even if not a single word survives revision.

A University of Florida photography professor tried an experiment that proves why this works. He divided his ceramics class into two groups. One group would be graded solely on quantity—pounds of clay shaped into pots at semester's end determined the grade. The other group would be graded on quality—they needed to produce only a single pot, as close to perfect as possible.

At semester's end, something surprising emerged. The highest-quality pots all came from the quantity group. While they'd been churning out pot after pot, they'd been learning from mistakes. They'd developed skill through volume. The quality group had spent the semester theorizing about perfection without practicing enough to achieve it.

This research, documented in *Art & Fear*, reveals something crucial about creative development: quality emerges from quantity. Skill develops through repetition, not through careful planning of a single attempt. AI makes you a quantity-group creator. When generating ten headlines takes ten minutes instead of two hours, you can afford to throw away nine and keep the one that works. The mediocre outputs aren't waste—they're the necessary raw material for identifying excellence.

Iterative Learning

The professionals who struggle with AI treat it like a search engine. They type a query, examine the result, and either accept or reject what they receive. One shot. Pass or fail.

This approach wastes most of AI's value.

AI interaction works better as dialogue. The first response is rarely optimal—but it's information about what the AI understood, which becomes material for refinement. "That's too formal" leads to an adjusted tone. "Focus more on the cost implications" redirects emphasis. "Give me five alternatives to that opening" generates options. Each exchange builds toward better output.

Think of it as collaborative refinement, not request-and-receive. You're working with the AI to develop something useful, not ordering from a menu and hoping you like what arrives.

This iterative approach requires patience initially but produces far better results. The skills transfer—effective prompts in one domain often work in others. You develop general principles for AI collaboration that compound over time.

Managing Frustration

The AI will fail you. Count on it.

It will generate plausible-sounding information that turns out to be fabricated. It will miss your point entirely and answer a question you didn't ask. It will produce the same unhelpful response three times despite your attempts to redirect. It will insert awkward phrases no human would ever use.

These failures are not reasons to abandon AI tools. They're reasons to develop robust verification habits and realistic expectations.

When AI fails, the productive response is curiosity, not frustration. Why did this fail? Was my instruction unclear? Is this the wrong tool for this task? What different approach might produce better results?

Professionals who integrate AI successfully treat failures as data points. Each failure teaches something about the tool's limitations, about their own prompting patterns, about which tasks are genuinely suited to AI assistance. The professionals who give up after early frustrations miss the learning that leads to reliable value.

Expect failures. Build verification into your process. Pivot strategies when something isn't working. The goal isn't perfect AI—it's AI that reliably improves your practice despite its imperfections.

THE ONE-EMAIL CHALLENGE

Here's a challenge that costs almost nothing: for the next five working days, delegate exactly one routine email per day to AI.

Not your whole correspondence. Not anything high-stakes. One email, daily, that you would normally write yourself. A scheduling request. A meeting follow-up. A status update. Something that takes you five minutes normally.

The constraint is deliberate. One email is small enough that failure is acceptable. If the AI output is unusable, you've lost five minutes. If you don't like how it feels, you can stop after day one. The commitment is minimal enough that procrastination has nothing to grip.

But one email is also enough to learn something real. Over five days, you'll encounter what AI does well and what it misses. You'll develop prompting intuitions without reading a single guide. You'll experience the editing cycle—reading output, noting deficiencies, requesting changes, evaluating improvement. You'll have data about how this fits your work specifically.

After five days, you have three options. Stop—the experiment wasn't worth it, and you've invested less than an hour total. Continue at the same pace—one email daily is genuinely useful, and you're comfortable with that level. Or expand—the experiment revealed possibilities, and you're ready to try more ambitious delegations.

Most people discover they're ready to expand. Not because they were persuaded by arguments, but because they experienced results. The challenge works precisely because it's small enough to try without conviction and revealing enough to build conviction through experience.

Start tomorrow. One email.

YOUR FIRST WEEK OF FREEDOM

If everything in this chapter feels like a lot—too many concepts, too many considerations, too many nuances—here's the simplified version.

Your first week of AI integration requires exactly five actions. Complete them in order. Don't complicate beyond what's specified.

Day 1: Choose Your Tool (15 minutes)

Open a browser. Navigate to either chat.openai.com (ChatGPT) or claude.ai (Claude). Both offer free tiers. Pick one. Don't research which is better—that distinction matters far less than actually starting. Create an account. Your tool selection is complete.

The common mistake: spending an hour reading reviews or comparison articles. Resist. Any major AI assistant will serve for beginner experiments. You can switch later if you discover reasons to.

Day 2: Find Your First Delegation Candidate (10 minutes)

Open your sent email folder. Scan the past week's correspondence. Find one message that meets these criteria: you write similar messages regularly, the content didn't require strategic thinking, and it took longer than it was worth.

Candidates include: meeting scheduling requests, status updates to stakeholders, follow-up thank-you notes, information requests, routine confirmations. Copy the email text somewhere you can reference it.

The common mistake: choosing something too important. Your first delegation should be something you wouldn't mind redoing if the experiment fails. Low stakes, high repetition.

Day 3: Run the Experiment (20 minutes)

Open your AI tool. Describe what you need:

"I need to write a professional email about [topic]. The recipient is [role/relationship]. My goal is [specific outcome]. The tone should be [formal/casual/friendly/direct]. Here's the context: [relevant background]. Draft this email."

Read the output. Don't evaluate whether it's "good enough" yet. Just notice: what did the AI include? What did it miss? How does the structure compare to what you would have written?

Now iterate. Tell the AI what to fix: "Make the opening less formal." "Add a sentence about the timeline." "Shorten the closing." Request one change. See the result.

The common mistake: expecting first-try perfection. The goal of day 3 isn't a usable email—it's experiencing how AI responds to direction.

Day 4: Produce Something Usable (25 minutes)

Return to your tool. This time, iterate until the output is genuinely usable—something you could actually send. It might take three exchanges. It might take seven. Keep refining until the email meets your standard.

Note how long this took. Note how the final product compares to what you'd have written manually. Note which aspects required your input and which the AI handled adequately alone.

The common mistake: being too forgiving or too harsh. The standard is "would I send this?"—not "is this perfect?" and not "is this minimally acceptable?"

Day 5: Repeat and Reflect (15 minutes)

Find another email in the same category. Run the experiment again. This time, you'll be faster—you know what prompts work, what refinements to request, what the tool handles well.

After producing the usable output, spend five minutes reflecting: What did I learn about AI this week? What surprised me? What tasks might this apply to? Write down three observations.

The common mistake: stopping after a single experiment. One success might be luck. Repetition builds confidence that the approach works reliably.

What You've Accomplished

After this five-day sequence, you've done what most professionals never do: you've actually used AI for professional work and formed your own conclusions about its value. You have experience instead of theory. You have specific observations about what works in your context instead of generic advice.

You haven't mastered AI—that takes months. But you've broken the paralysis of not starting. You've proven to yourself that the tool responds to your direction. You've taken the first step toward the transformation this book promises.

Everything else builds on this foundation. Next week, expand to a different task category. The week after, try a longer project. The habits you build this week compound over time. But they only compound if you actually start.

Five days. Roughly ninety minutes total. That's the entire investment required to move from "thinking about AI" to "using AI." Begin tomorrow.

PREPARING FOR THE RENAISSANCE

You've addressed the psychological barriers. You've established ethical guidelines. You've adopted an experimental mindset and have a concrete plan for your first week. One task remains before action begins: preparing the conditions for successful implementation.

The temptation is to evaluate everything. With over 14,000 AI tools launched in 2023 alone—nearly forty new tools every day—surely the perfect solution is out there, waiting to be discovered through careful comparison.

This approach guarantees paralysis.

You don't need the optimal tool. You need a functional tool you'll actually use. The difference between the best AI writing assistant and the second-best is marginal compared to the difference between using any capable tool and using nothing.

Start with one text-based AI tool—ChatGPT, Claude, or a similar established option. These platforms handle most professional use cases adequately: drafting, summarization, brainstorming, analysis, research assistance. Learn one well before adding others.

If your work involves visual content, add one image generation tool. If it doesn't, skip this category entirely. You're building a minimal toolkit, not a comprehensive collection.

Ignore the hype cycle. Every week brings breathless announcements about new capabilities, new models, new platforms that will change everything. Most of these announcements are noise. The fundamentals of AI assistance don't change with each product launch —your skills with existing tools remain valuable even as the landscape evolves.

Evaluate new tools only when you encounter a specific limitation with your current setup. "I can't do X with my current tools, and X matters enough to justify the time investment of learning something new." This constraint-based approach prevents endless tool shopping while ensuring you adopt genuinely useful additions.

Defining Your Sacred Tasks

Not everything should be delegated. Part of strategic AI integration is explicitly identifying what you will never outsource—the work that remains human-only because the work itself is part of your value proposition.

These boundaries are personal and professional. They depend on your role, your industry, your sense of what matters.

Some candidates for protection: genuine relationship building, high-stakes conflict resolution, creative vision that defines your professional signature, strategic decisions that shape organizational direction, feedback and mentorship that requires authentic human presence.

Some tasks might feel sacred but actually aren't: routine correspondence that follows predictable patterns, first drafts that will undergo heavy revision anyway, research gathering that precedes your actual analysis, formatting and organization that serve primarily functional purposes.

The point isn't to minimize AI assistance. It's to make deliberate choices about where AI belongs and where it doesn't—choices based on your values and professional identity rather than default habits or vague discomfort.

Defining these boundaries reduces anxiety. When you know what you're protecting, you can delegate everything else without existential uncertainty about whether you're compromising something essential. The sacred tasks remain sacred. Everything else becomes fair game for strategic automation.

YOUR PERSONAL CONTRACT

This book asks for a commitment.

Not a vague intention to "use AI more." Not a general resolution to "stay current with technology." A specific commitment to engage with this material actively for sixty days.

The commitment involves weekly experiments—dedicated time to practice AI collaboration in the domains covered by each chapter. It involves daily micro-practices—small interactions that build fluency without requiring significant time investment. It involves honest evaluation of what works and what doesn't, captured in a format you can review and learn from.

In return, this book offers a structured path from where you are now—capable but uncertain, interested but hesitant—to somewhere genuinely different. To professional practice that integrates AI strategically. To work that focuses your energy on uniquely human contributions. To have confidence in your ability to direct technology rather than be threatened by it.

The next chapter introduces the framework that structures this journey: the five-step Professional Renaissance Method. Release, Expand, Partner, Protect, Elevate—each step builds on what you've learned here about overcoming resistance and establishing ethical foundations. But the framework is structure, not substance. The substance comes from your willingness to experiment, your honesty about results, and your commitment to the sixty-day transformation.

You've diagnosed the problem. You've overcome the internal resistance. You've established the ethical guidelines that allow you to proceed with integrity. You've prepared the conditions for success. You have a concrete plan for your first week.

Now the real work begins.

3

THE RENAISSANCE

"There is nothing so useless as doing efficiently that which should not be done at all." — Peter Drucker

EFFICIENCY CAN DESTROY YOU. Not through failure or collapse. Through quieter erosion that professionals rarely notice until months have passed. You automate tedious tasks, save two hours daily, and feel accomplished. Six months later, those two hours simply vanished into expanded versions of everything else. More emails. More meetings. More busy work wearing different clothing.

AI efficiency without strategic direction creates speed without destination.

The 5-Pillar Method described in this book exists to prevent this outcome. Not by making you faster at what you already do. By fundamentally changing what you spend your time doing.

The pattern is predictable. Professionals encounter compelling ideas about AI integration, feel motivated for a few days, try experiments without structure, encounter friction, and gradually return to old habits. Six months later, they're reading another article about the same topic, feeling the same mixture of interest and guilt.

The 5-Pillar Renaissance Framework exists to break this pattern. Not through motivation—motivation fades. Through structure that makes progress systematic and sustainable.

Most systems fail for a predictable reason: they ask you to become someone else. Wake earlier. Track everything. Develop habits that feel foreign to your natural rhythms. Initial enthusiasm carries you for days, maybe weeks. Then real life intervenes—a crisis at work, a family obligation, an energy dip—and the system shatters because it requires conditions that reality doesn't reliably provide.

This method is designed differently. It doesn't ask you to transform your personality. It asks you to shift how you allocate attention you already have.

Consider what the framework actually requests: thirty minutes daily of focused experimentation, a weekly review of what you tried. That's it. Not a morning routine revolution. Not a complete calendar restructuring. Not the kind of heroic discipline that sounds inspiring in books and proves impossible in practice.

The experimentation itself should feel engaging, not effortful. You're not doing push-ups. You're exploring whether AI can handle tasks you find tedious. Most people find this genuinely interesting once they start—watching a machine struggle with something they've done a thousand times, then suddenly produce something usable.

Try it as a real experiment, not a commitment to a new identity. Scientists don't feel like failures when hypotheses prove wrong. They feel informed. That's the stance to take here.

THE 5-STEP FRAMEWORK OVERVIEW

The Professional Renaissance Method consists of five sequential steps, each building on the one before. The sequence matters. Professionals who skip ahead—jumping straight to creative AI applications without first clearing their routine workload—typically struggle to sustain practice. The cognitive bandwidth isn't available. The time isn't available. The foundation isn't in place.

The five steps are Release, Expand, Partner, Protect, and Elevate. Each represents both a practical domain of AI integration and a mindset shift that enables that integration.

Step 1: Release

Release addresses the Robot Zone work identified in your Cognitive Energy Audit—the repetitive, predictable tasks that consume time without requiring your particular expertise. Scheduling, routine correspondence, basic research gathering, document formatting, status updates that follow predictable patterns.

The mindset of Release is delegation without guilt. This requires letting go of ego attachment to busy work—the subtle pride in being the person who handles everything personally. Release recognizes that time spent on tasks any competent professional could perform is time unavailable for work that actually requires your judgment and experience.

The practical goal of Release: systematically identify and transfer routine tasks to AI assistance, reclaiming hours currently lost to work beneath your capability level. Most professionals who complete this step thoroughly recover five to ten hours weekly. Not through working faster. Through working on different tasks entirely.

Step 2: Expand

Expand applies AI to creative and generative work—ideation, brainstorming, option generation, draft creation. This is where AI shifts from efficiency tool to amplification partner.

The mindset of Expand is abundance rather than scarcity. Traditional creative work operates under constraints: you generate the ideas you can generate, limited by your individual cognitive capacity and the time available for exploration. AI removes this constraint. You can generate fifty options to find the three worth developing. You can explore ten different angles before committing to one. You can produce rapid prototypes to test whether concepts have potential.

The practical goal of Expand: multiply your creative output by using AI to generate options, variations, and starting points that your expertise then curates and refines. Professionals in this phase often report that work which previously felt like extraction—pulling ideas from an increasingly depleted well—begins to feel like selection among genuine alternatives.

Step 3: Partner

Partner addresses analytical work—research synthesis, data interpretation, pattern recognition, scenario modeling. This is collaborative cognition: human contextual understanding combined with AI processing capacity.

The mindset of Partner is collaborative sense-making. You're not outsourcing analysis. You're conducting analysis with an assistant that can hold more information simultaneously than your working memory allows, surface patterns you might miss in complex datasets, and generate alternative interpretations you might not consider independently.

The practical goal of Partner: establish workflows where AI handles information processing while you provide context, judgment, and interpretation. The AI summarizes the fifty-page report. You determine what the summary means for your specific situation. The AI identifies trends in the data. You assess whether those trends reflect reality or artifacts of how the data was collected.

Step 4: Protect

Protect identifies and safeguards work that should remain exclusively human. This step might seem counterintuitive in a book about AI integration. It's essential to sustainable practice.

The mindset of Protect is intentional humanity. Some professional activities derive value precisely from human presence, human judgment, and human connection. Automating these activities doesn't create efficiency—it destroys value. A mentor who outsources feedback to AI isn't being efficient. They're abandoning the relationship. A leader who automates difficult conversations isn't saving time. They're eroding trust.

The practical goal of Protect: explicitly define your "Human-Only Zone"—the interactions, relationships, and decisions where authentic human presence is non-negotiable. This isn't about technical limitations. It's about what your professional identity requires and what your stakeholders deserve.

Step 5: Elevate

Elevate applies AI to strategic thinking—decision-making, planning, assumption-testing, long-term visioning. This is the highest-leverage application: using AI not to do your thinking, but to improve the quality of your thinking.

The mindset of Elevate is intellectual expansion. AI becomes a sparring partner that challenges your assumptions, generates counter-arguments, stress-tests your strategies, and surfaces considerations you might have missed. You remain the decision-maker. The decisions become better informed.

The practical goal of Elevate: develop practices for AI-assisted strategic thinking that improve decision quality without surrendering decision authority. The AI generates five distinct strategic options. You choose among them based on values, relationships, and context the AI cannot fully assess. The AI identifies weaknesses in your plan. You determine which weaknesses matter enough to address.

THE SEQUENTIAL LOGIC

The sequence isn't arbitrary. Release creates time for Expand. Expand builds creative confidence for Partner. Partner develops evaluation skills for Protect. Protect clarifies boundaries that enable Elevate.

Dr. Sophie Leroy's research on "attention residue" demonstrates that switching between tasks leaves cognitive traces—part of your mind remains on the previous task even as you engage the new one. When routine work remains unaddressed, it creates a persistent residue that degrades performance on complex thinking. You can't think strategically with half your cognitive bandwidth occupied by administrative overhead. Release clears the residue before Elevate demands full attention.

Professionals who attempt to start with Elevate—jumping directly to AI-assisted strategic thinking—often struggle because they haven't built the foundational skills. They can't effectively evaluate AI output because they haven't practiced evaluation on lower-stakes tasks. They lack the time for strategic thinking because routine work still consumes their weeks. They haven't developed the judgment about when AI adds value and when it doesn't.

The temptation to skip ahead is understandable. Strategic AI thinking sounds more important than administrative cleanup. Why spend two weeks on emails when you could be red-teaming business strategy?

Because cognitive bandwidth is finite. Occupied bandwidth isn't available for new work. When your routine tasks remain unaddressed, they consume mental background processes. The unanswered emails create ambient anxiety. The unfinished reports generate cognitive weight. The upcoming meeting that needs prep occupies a thread of attention. This background load leaves less capacity for focused experimentation with new AI capabilities.

More practically: the time isn't available. Professionals who try to add strategic AI practices without first clearing routine work discover they can't sustain the addition. The new practice competes with existing obligations rather than replacing them. Within weeks, the new practice loses because it's newer—the established patterns have institutional weight.

Release creates space before Expand fills it. Partner builds evaluation skills that Protect requires. Each step genuinely enables the next. Not arbitrary sequencing for aesthetic purposes. Load-bearing structure that makes the later steps achievable.

Professionals who insist on skipping typically return weeks or months later, frustrated that their AI efforts haven't compounded. When they actually complete Release, they discover the missing foundation. Everything else becomes easier because they're no longer trying to build on air.

Start where the framework says to start. The sequence exists because it works.

THE "AMPLIFY" PHILOSOPHY

The five steps share an underlying philosophy that distinguishes strategic AI integration from mere tool adoption. Understanding this philosophy prevents the most common failure mode: using AI to become average faster.

The obvious pitch for AI tools focuses on efficiency. Do more in less time. Automate the tedious. Accelerate your workflow.

This pitch is accurate but incomplete. When it becomes the primary frame, it leads professionals astray.

Efficiency gains without strategic direction create a specific trap: Parkinson's Law. Work expands to fill available time. If you save two hours daily through AI automation but don't intentionally direct those hours toward higher-value work, the time simply disappears into expanded versions of whatever fills your days. More emails. More meetings. More busy work dressed in slightly different clothing.

The goal isn't efficiency for its own sake. The goal is cognitive freshness for work that matters and quality improvement in the outputs that define your professional reputation.

A two-hour task that becomes a twenty-minute task is valuable not because you completed it faster, but because you now have one hundred minutes available for work that requires concentrated attention and fresh thinking. The value is in the reallocation, not the acceleration.

THE DIRECTOR'S CHAIR

Film production involves hundreds of specialists: cinematographers, editors, sound designers, lighting technicians, costume designers, set builders. Each brings technical expertise the director likely doesn't possess. Yet the film remains the director's vision.

The director doesn't operate the camera or sew the costumes or build the sets. The director ensures that all these contributions serve a coherent artistic purpose. Technical execution belongs to specialists. Vision, integration, and quality standards belong to the director.

Consider what this actually means on set. The cinematographer focuses on the camera—on lighting, angles, movement through space. The actor focuses on their character—on emotional truth, on dialogue, on the scene partner's reactions. The editor focuses on pacing—on which take works, on when to cut, on how sequences flow. Each specialist does essential work. But each sees the project from their particular vantage point.

Only the director holds the complete vision: how this shot serves this scene, how this scene serves this sequence, how this sequence serves the entire film. Legendary directors build epic careers in exactly this capacity—the ability to see the whole picture while everyone else is necessarily focused on their part of it.

This model applies directly to AI-augmented professional practice. You're becoming the director of your work.

AI handles technical execution—generating drafts, processing information, producing options. You provide vision, context, and quality judgment. You determine what the work needs to accomplish. You evaluate whether the output serves that purpose. You integrate individual contributions into coherent outcomes.

This doesn't mean delegation of responsibility. It's an elevation of role. The director isn't less valuable because they don't personally operate every piece of equipment. They're more valuable because their expertise focuses where it matters most—on the decisions that determine whether the final product succeeds.

Your AI tools are specialists. Extremely capable within their domains, tireless, fast—but fundamentally focused on the immediate task. You're the one who sees the whole picture. The one who knows what this particular output needs to accomplish, who it needs to serve, what context the AI doesn't have.

COGNITIVE LEVERAGE

Working memory—the mental workspace where you hold and manipulate information—has hard limits. George Miller's famous 1956 paper established working memory's limit at "seven plus or minus two" items. More recent work by Nelson Cowan suggests the true limit is closer to four chunks of information.

Either way, complex strategic decisions routinely involve more variables than your brain can consciously juggle.

AI provides cognitive scaffolding that extends these limits. Not by thinking for you. By holding information while you think.

You can ask AI to hold ten variables while you focus on the relationships between three. You can generate multiple scenarios and compare them side by side. You can access synthesized information from sources too numerous to read directly.

This leverage doesn't replace your thinking. It expands the territory your thinking can cover. You're still providing direction, judgment, and interpretation. You're doing it with access to more information, more options, and more explicit consideration of factors that might otherwise remain implicit.

The professional who uses AI for cognitive leverage doesn't become dependent on AI for thinking. They become capable of thinking about bigger problems, more complex situations, longer time horizons—the thinking that actually distinguishes strategic contribution from tactical execution.

THE ATROPHY QUESTION

A legitimate concern accompanies these arguments: will relying on AI for certain tasks cause your underlying skills to deteriorate?

The concern isn't irrational. Skills require practice. If AI drafts all your emails, will you lose the ability to write emails yourself? If AI handles your research synthesis, will your analytical capacity atrophy?

This fear deserves the most direct answer because it carries the most emotional weight.

The answer depends on which skills we're discussing—and on a distinction most atrophy arguments miss entirely.

Professional expertise operates on two levels that are frequently confused: what you can do and what you are doing. Capability is your accumulated understanding—the mental models, the pattern recognition, the judgment developed over years of practice. Demonstration is the active execution—the work product your hands produce, the visible output of your effort.

These aren't the same. Conflating them leads to significant errors in thinking about AI integration.

When you delegate a task to AI, your demonstration changes—you're no longer producing that work manually. But your capability doesn't diminish. You still understand how the work should be done, what quality looks like, and what errors to catch. The capability transfers to supervision, direction, and evaluation rather than execution.

Consider the surgeon who adopts robotic assistance for certain procedures. Their hands no longer perform the precise movements the robot now handles. Are they losing surgical skill? Not in any meaningful sense. They still understand anatomy. They still make diagnostic judgments. They still recognize when something goes wrong and need to intervene. The capability remains. Only the demonstration shifts.

Research on robotic surgery demonstrates that surgeons maintain decision-making expertise even as physical execution shifts to machines. Surgeon judgment—selecting appropriate interventions, recognizing complications, making real-time adjustments—remains the primary determinant of patient outcomes regardless of whether procedures were performed manually or robotically. The expertise isn't in the hand movements. It's in the thinking that guides them.

What would cause actual skill atrophy is stepping away entirely—no longer thinking about the work, no longer evaluating output, no longer making judgments about quality. A surgeon who stopped operating would indeed lose surgical skill. But that's not what's happening here.

When you direct AI toward a writing task, you're not stepping away from writing. You're writing differently. You're evaluating the draft, identifying what doesn't work, refining the language, adding the elements that require your particular voice. The skill is in the evaluation and direction, not in the keystroke production.

The professional who has written ten thousand emails hasn't lost communication capability by delegating routine emails. They've freed that capability for communications that actually require their full attention. The analyst who has built five hundred spreadsheets hasn't lost analytical capability by using AI for routine data processing. They've redirected that capability toward interpretation that requires human context.

Capability atrophies only through complete disengagement—when you stop thinking about the work entirely, stop evaluating outputs, stop exercising judgment. That's not what happens in effective AI integration. Judgment remains in constant use. It's just applied differently.

There's also the question of which skills actually deserve preservation.

Your ability to write emails didn't develop because you planned to spend your career writing emails. It developed as a byproduct of learning to communicate professionally. The underlying capability—clear professional communication—isn't threatened by automating the routine execution. The capability transfers to higher-stakes communication that still requires your attention.

Similarly, your research skills won't atrophy because AI handles initial information gathering. You're still synthesizing, evaluating, and interpreting. The mental muscles that matter remain in constant use—just applied to judgment rather than retrieval.

The risk of atrophy is real for skills actively under development. Beginners should build foundations before relying on AI assistance. The risk is minimal for professionals with established capabilities who are shifting from execution to direction. You're not forgetting how to do the work. You're choosing where to apply your ability to do it.

Watch for the distinction in your own thinking. When you feel anxiety about "losing skills," ask: Am I actually at risk of losing capability, or just changing how I demonstrate it? The answer is almost always the latter.

The professionals who should worry about atrophy are those using AI as a crutch—accepting output without evaluation, delegating judgment rather than execution, stopping their engagement before the work is truly finished. If you're doing that, the framework you're about to learn will correct it. Verification is built in. Human judgment at every stage is built in.

Your skills evolve. They don't atrophy. You're becoming something more than what you were, not less.

THE 60-DAY JOURNEY ROADMAP

The Professional Renaissance Method unfolds across sixty days, divided into four phases. Each phase corresponds to specific steps in the framework and includes weekly experiments and daily micro-practices.

The sixty-day timeframe isn't a guess. Research by psychologist Phillippa Lally and colleagues at University College London tracked participants forming new habits and found that automaticity—the point where a behavior becomes effortless—developed between 18 and 254 days, with a median of 66 days. Simple habits are automated quickly. Complex behaviors involving multiple decisions took significantly longer. Sixty days provides a realistic runway for the interconnected habits AI integration requires.

Sixty days is also enough to maintain focus. You're not committing to an indefinite transformation project. You're committing to a defined period of intensive experimentation. The structure matters psychologically. Open-ended commitments ("I'm going to get better at AI") produce less action than bounded ones ("I'm running a sixty-day experiment"). Bounded commitments allow you to fully engage because you can see the finish line. They create urgency without panic.

If sixty days still feels too long, try this: commit to the first fourteen days—Phase 1, the Release phase. That's enough time to see tangible results. If those results don't justify continuing, stop. But if they do—if you actually reclaim hours and feel the difference—the remaining forty-six days will feel less like obligation and more like exploration you want to continue.

Phase 1: The Clean Up (Days 1-14)

Phase 1 implements Release. The focus: clearing the deck of administrative debris.

During these two weeks, you'll systematically audit your routine tasks, identify the highest-value delegation candidates, and begin transferring Robot Zone work to AI assistance. The goal isn't comprehensive automation. The goal is momentum—tangible results that demonstrate AI's practical value in your specific context.

Week 1 focus: Complete or update your Cognitive Energy Audit. Identify your three to five most time-consuming routine tasks. Select one task for immediate AI delegation. Experiment with transferring that task, documenting what works and what requires adjustment.

Week 2 focus: Extend AI delegation to additional routine tasks. Begin building a prompt library—saved instructions for tasks you'll delegate repeatedly. Establish quality verification habits. Measure time reclaimed.

By Day 14, you should have clear evidence of practical value: hours reclaimed through successful delegation, a small library of effective prompts, and growing confidence in evaluating AI output for routine tasks.

Here's what changes after completing just the Release phase—two weeks into the framework.

Valerie, a marketing director, previously spent her mornings on email triage, meeting scheduling, and status report compilation. By

10:30 AM, she'd done nothing that actually required her expertise but felt exhausted from constant task-switching.

Post-Release morning: She reviews AI-drafted responses to routine emails in ten minutes, approving most with minor edits. Scheduling requests have been handled automatically—she just confirms the calendar updates. The weekly status report was compiled overnight from project management data. She reads it, adds one insight about team morale, and sends it.

By 9:30 AM, she's completed work that used to consume her until 10:30. And she's not tired yet—none of that work required creative or strategic thinking, so her cognitive reserves are intact.

She uses the next ninety minutes for campaign strategy—work she used to attempt at 4 PM when her brain was fried. The ideas come easier. The decisions feel clearer. She can actually think about the quarter ahead rather than reacting to the day's demands.

Her team notices she's more present in meetings. She's not checking her phone for the next urgent message because the urgent messages are being screened. She's not anxiously calculating whether she'll finish her report because the report is already done.

The afternoon contains the same meetings as before, but she arrives at them differently. There's still stress—the marketing budget conversation at 3 PM is legitimately difficult. But it's the only stress, not one stress layered on top of background administrative anxiety.

By 5:30 PM, she's done. Actually done, not just temporarily pausing the flood. Tomorrow's routine work is already queued for AI processing. She leaves without the laptop weight of unfinished tasks.

This is Release. Not magic, but meaningful.

Phase 2: The Creative Spark (Days 15-30)

Phase 2 implements Expand. The focus: ideation and overcoming creative blocks.

During these two weeks, you'll apply AI to generative and creative work—brainstorming, option generation, draft creation, concept development. The goal is expanding your creative capacity, not outsourcing creativity itself. You'll generate more options than you could alone, then apply your expertise to curate and refine.

Week 3 focus: Identify one creative project currently in progress or upcoming. Use AI to generate multiple approaches, outlines, or initial concepts. Practice the Draft Zero method—using AI to produce starting material your judgment then improves. Compare AI-generated options with your own initial instincts.

Week 4 focus: Extend AI collaboration to additional creative tasks. Experiment with persona prompts—asking AI to approach problems from different perspectives. Practice voice calibration—feeding AI examples of your work to improve stylistic alignment. Develop your approach to curation: selecting among generated options, combining elements, rejecting what doesn't fit.

By Day 30, you should have experience using AI as a creative amplifier, evidence that AI-assisted ideation produces viable options, and techniques for maintaining your creative voice while leveraging AI's generative capacity.

Phase 3: The Analytical Deep Dive (Days 31-45)

Phase 3 implements Partner. The focus: data, research, and synthesis.

During these two weeks, you'll develop collaborative analysis workflows. AI handles information processing—summarization, pattern identification, scenario generation. You provide context, judgment, and interpretation. The combination produces insights neither could achieve alone.

Week 5 focus: Select an analytical challenge—a report to digest, data to interpret, or research question to explore. Experiment with AI summarization capabilities. Ask AI to identify patterns, gaps, or tensions in the material. Practice verification: checking AI analysis against your own reading and against source documents.

Week 6 focus: Extend collaborative analysis to additional projects. Experiment with scenario modeling—asking AI to project multiple possible outcomes. Practice Socratic dialogue with AI—using questions to probe documents, test assumptions, and surface considerations you might miss. Develop your approach to synthesis: combining AI-generated analysis with contextual knowledge AI doesn't possess.

By Day 45, you should have functional workflows for AI-assisted analysis, confidence in your ability to verify and interpret AI-generated insights, and evidence of improved analytical capacity through human-AI collaboration.

Phase 4: The Human Reconnection (Days 46-60)

Phase 4 implements both Protect and Elevate. The focus: relationships and high-level strategy.

During these final two weeks, you'll accomplish two objectives. First, you'll explicitly define your Human-Only Zone—the relationships, interactions, and decisions where AI has no place. Second, you'll develop practices for AI-assisted strategic thinking—using AI to improve decision quality without surrendering decision authority.

Week 7 focus: Conduct a relationship audit. Identify the interactions and touchpoints that require authentic human presence. Explicitly document your boundaries: where AI will never be involved. Reinvest time reclaimed from earlier phases into relationship deepening. Experiment with AI as relationship support—preparing for meetings, remembering context—while keeping the human interaction itself AI-free.

Week 8 focus: Develop strategic sparring practices. Use AI to critique your plans, generate counterarguments, and surface blind spots. Practice the pre-mortem technique—asking AI to describe how your strategy might fail, then working backward to address vulnerabilities. Distinguish between AI-generated input (valuable) and AI-directed decisions (inappropriate). Document your emerging approach to AI-augmented strategic thinking.

By Day 60, you should have clear boundaries protecting your most valuable human contributions, practices for AI-assisted strategy that improve decision quality, and an integrated approach to AI collaboration that spans routine work through high-stakes thinking.

SETTING THE STAGE FOR SUCCESS

Structure increases the probability of success. Before beginning the sixty-day journey, establish the conditions that make sustained practice possible.

Required Resources

The technical requirements are deliberately minimal.

You need one capable text-based AI tool. ChatGPT Plus, Claude Pro, or a similar established platform. Any of these will handle the applications covered in this book. Don't spend time comparison shopping—choose one, learn it well, and add tools only when you encounter specific limitations that matter.

You need a system for capturing what you learn. A note-taking application, a physical notebook, a document folder—anything that allows you to record prompts that work, approaches that fail, insights that emerge from experimentation. This documentation is valuable. Memories fade, but written records compound.

If your work involves visual content, add one image generation tool after mastering the text tool. If your work doesn't involve visual content, skip image generation entirely. You're building a minimal effective toolkit, not collecting subscriptions.

Resist the impulse to acquire every AI tool that seems potentially useful. Each tool requires learning time. Each tool fragments your attention. The professionals who integrate AI successfully typically master a small number of tools deeply rather than accumulating a large number of tools superficially.

Time Blocking

Daily practice matters more than session length. Thirty minutes daily produces better results than three hours weekly.

Block thirty minutes each day for AI experimentation. Protect this time as you would protect a meeting with your most important client.

The experiments don't require more time than this—but they do require consistency. Skills develop through repeated practice. Habits form through daily repetition. Insights accumulate through sustained attention.

Place the thirty-minute block where it's most likely to actually happen. For some professionals, early morning before the reactive demands begin. For others, late afternoon when routine work is winding down. The timing matters less than the consistency.

If thirty minutes feels impossible, start with fifteen. A shorter block practiced daily beats a longer block practiced sporadically. You're building a sustainable practice, not performing a temporary sprint.

Weekly Reviews

Each Friday, spend fifteen to thirty minutes reviewing the week's experiments. This review serves multiple functions.

Documentation: Record what you tried, what worked, what didn't. Capture prompts that produced useful output. Note approaches that failed and your hypotheses about why.

Assessment: Evaluate progress against the phase goals. Are you on track? What obstacles have emerged? What adjustments would improve next week?

Adjustment: Update your approach based on evidence. Double down on experiments showing promise. Abandon or modify experiments showing poor results. Identify the next experiment to prioritize.

The Friday review transforms random experimentation into systematic learning. Without it, you'll struggle to remember what you tried, miss patterns in what's working, and repeat mistakes you've already made.

Measuring Success

Traditional productivity metrics—tasks completed, hours worked, output volume—miss what matters most in AI integration. Speed improvements are real, but they're means, not ends.

More useful questions for evaluation:

Did I spend more time on strategic work this week than last week? The goal is shifting time toward higher-value contributions, not just completing existing work faster.

Did I produce work of higher quality? AI should improve outputs, not just accelerate them. Better ideas. Sharper analysis. More considered decisions.

Did I feel more cognitively fresh? Sustainable practice requires energy management. If AI integration leaves you more drained—through the effort of learning new tools, the stress of unfamiliar processes—something needs adjustment.

Did I protect the relationships and interactions that matter most? Efficiency that damages relationships isn't success. Success means reinvesting reclaimed time into human connection, not eliminating human connection in pursuit of optimization.

Track these questions weekly in your review. Look for patterns over the sixty days. The trajectory matters more than any single week's results.

THE 60-DAY CONTRACT

Before beginning the framework, complete this brief commitment ritual. It takes less than five minutes but significantly increases follow-through.

Step 1: Define your primary motivation. Complete this sentence in writing: "I am undertaking the Professional Renaissance Method because I want to ____________." Be specific. Not "use AI better" but "reclaim Wednesday afternoons for deep work" or "stop dreading the monthly reports."

Step 2: Identify your accountability mechanism. Who will you tell about this commitment? A colleague, friend, or family member who will ask how it's going. Social commitment increases completion rates dramatically.

Step 3: Anticipate your obstacle. Complete this sentence: "The most likely thing to derail my commitment is ____________." Naming the obstacle in advance reduces its power when it arrives. If you know travel will interrupt Week 3, you've already mentally prepared to resume afterward rather than abandon the effort.

Step 4: Choose your non-negotiable minimum. What will you do on the days when life intervenes—when thirty minutes is genuinely impossible? Perhaps: "I will at least log one observation about AI potential in my notes." The minimum keeps the chain unbroken even when circumstances strain.

Step 5: Sign and date. This feels ritualistic because it is. Written commitments activate psychological consistency mechanisms that verbal intentions don't. Put your name on this.

You're not signing a contract with me or with any external authority. You're signing a contract with the version of yourself who will exist sixty days from now—more capable, more strategic, more valuable. That person is counting on present-you to follow through.

You have the framework. You have the philosophy. You have the roadmap. You have the setup for success.

The next chapter begins implementation. Step 1: Release. You'll learn exactly how to identify your highest-value delegation candidates, construct effective AI prompts for routine tasks, and build the verification systems that ensure quality without micromanagement.

Theory becomes practice. The Renaissance begins.

4

RELEASE THE ROUTINE

"The difference between successful people and really successful people is that really successful people say no to almost everything." — Warren Buffett

TRANSFORMATION HAPPENS IN THE MARGINS. Not the visible work—the invisible recovery between concentrated efforts.

The person who produces breakthrough analysis at 2 PM does so partly because of cognitive skill and domain expertise. But also because they weren't exhausted from spending 9 AM to 1 PM on routine correspondence. The strategic insight that emerges during an afternoon walk happens because there was cognitive space for synthesis—space that wouldn't have existed if the morning had been consumed by administrative coordination.

This isn't about working less. It's about working strategically—deploying your attention where it compounds rather than depletes. Routine work depletes. It consumes energy and returns completion. Strategic work compounds. It builds capabilities, generates insights, creates opportunities that multiply over time.

The audit revealed how much of your energy goes toward depletion activities. This chapter redirects that energy toward compound activities. Not eventually. This week. With delegation protocols you can implement before Friday.

The transformation available through this single chapter exceeds what most professionals accomplish in a full year of conventional productivity improvement. Not because the techniques are revolutionary. Because the reallocation is fundamental: stop spending yourself on what depletes and start investing yourself in what compounds.

You may feel ready for tactics. But something still feels unresolved. The mindset work in Part I addressed the big-picture anxiety. Now, as you're about to actually delegate real work to AI, smaller concerns may start to surface—practical worries about what delegation means for your professional identity and competence.

Let's address them directly.

"My work is too nuanced for AI to handle properly."

This objection masks a deeper concern: if AI can handle your work, maybe your work wasn't as valuable as you believed.

Both layers deserve examination.

First, you're probably right that some of your work is too nuanced for AI. The high-stakes client communication that requires reading between the lines. The strategic analysis that draws on years of pattern recognition. The relationship repair that demands emotional intelligence and perfect timing. These tasks belong in your Protect zone, not your Release zone.

But not all of your work is that nuanced. That's the honest assessment this chapter asks you to make.

Look at your Cognitive Energy Audit. Somewhere in there are tasks that feel professional and important but follow predictable patterns. The meeting confirmation emails that vary only in dates and names. The status updates that use the same structure every week. The document formatting that applies consistent standards to different content. These aren't nuanced—they're routine work dressed in professional clothing.

The nuance concern conflates two different things: the importance of getting work right and the requirement for human execution. Important work and routine work aren't opposites. Some routine work is very important—and still routine. The meeting confirmation email matters. You need to get the time and location right. But "getting it right" doesn't require your creative judgment. It requires following a clear process correctly.

Here's the second layer: if some of your work turns out to be less nuanced than you assumed, that's not a threat to your professional value. It's an opportunity to redirect your energy toward work that actually is nuanced.

The professional who insists on personally handling every task—nuanced or not—isn't protecting their value. They're diluting it. When you spend three hours on work that AI could handle acceptably, you're not demonstrating nuanced capability. You're demonstrating poor time allocation.

Your nuanced work becomes more visible, more frequent, and more impactful when routine work isn't competing for the same hours. Delegation doesn't diminish your nuanced contributions. It creates space for them to matter more.

"Using AI for professional work feels like cheating."

You've been praised throughout your career for hard work. For putting in the hours. For doing the heavy lifting yourself. That conditioning doesn't disappear because technology changed.

The "cheating" feeling comes from conflating effort with value. We're taught that struggle validates outcomes—that the essay written at 2 AM is somehow more legitimate than one produced efficiently. This made sense when effort was the primary input to quality. It makes less sense when better tools are available.

Consider how this logic applies elsewhere. Is the architect who uses CAD software cheating compared to one who drafts by hand? Is the accountant who uses spreadsheets cheating compared to one who calculates manually? Is the writer who uses a word processor cheating compared to one who types on a mechanical typewriter?

At every technological transition, there's a period where using new tools feels like cutting corners. Then the new baseline gets established. Refusing the tools starts looking like professional stubbornness rather than integrity.

The "cheating" frame also misunderstands what you're actually providing. Your value isn't the keystrokes. It's the judgment that shapes what gets written, the standards that define what's acceptable, the expertise that catches errors and elevates quality. A routine email that you delegated but reviewed isn't less valuable than one you typed yourself—it delivers the same outcome while freeing your capacity for work that actually requires your expertise.

Nobody thinks the CEO is "cheating" when they have an assistant manage their calendar. The CEO's job is making high-quality decisions and building key relationships—not personally coordinating meetings. The assistant creates leverage. The executive's judgment is applied where it matters most.

You're becoming that CEO of your professional practice. Your judgment is scarce and valuable. The mechanical execution of routine tasks is now available at scale. Applying your judgment to more consequential work isn't cheating. It's professional maturity.

THE PERSONAL ASSISTANT FRAME

Think of what's about to happen as hiring your first personal assistant. Not a replacement for your judgment—an extension of your capacity.

A senior executive doesn't answer their own phone, manage their own calendar, or format their own documents. They have an assistant who handles these tasks—not because the executive is incapable, but because their time is worth more applied elsewhere. The assistant creates leverage. The executive's ideas reach more people, their decisions happen faster, their attention focuses on what matters most.

For decades, this leverage was available only to those at the top of organizational hierarchies. You needed budget, authority, and status to have someone managing your administrative burden. Everyone else did their own scheduling, their own formatting, their own routine correspondence—regardless of whether that was the best use of their capabilities.

AI has democratized the personal assistant.

You now have access to a tireless, patient, always-available assistant who can handle a specific category of tasks: the predictable, pattern-based, rules-following work that currently consumes hours of your week. This assistant doesn't need benefits, doesn't take sick days, and costs less per month than a single hour of professional consulting time.

But here's what makes this different from hiring a human assistant: you don't need to trust this assistant's judgment. In fact, you shouldn't. A good human assistant develops understanding of your preferences over time and eventually makes autonomous decisions on your behalf. The AI assistant doesn't have that trajectory. Every output requires your review. Every decision remains yours.

This is an advantage, not a limitation. You get the leverage of delegation without the vulnerability of trusting another person's judgment with your professional reputation. The assistant handles the mechanical work. You provide the oversight that ensures quality.

The frame matters because it shapes how you approach the protocols that follow. When you're "prompting AI," it can feel technical, impersonal, even desperate—like you're trying to trick a machine into doing your work. When you're "directing your assistant," you're doing what every effective leader does: communicating clearly what you need, providing appropriate context, and establishing standards for acceptable output.

The personal assistant frame also clarifies the emotional dynamic. People feel guilty about "making AI do their work," as if they're cheating or cutting corners. Nobody feels guilty about delegating scheduling to an assistant. That's what assistants are for. The guilt comes from misunderstanding the relationship.

Your AI assistant is not a replacement for your capabilities. It's a resource that makes your capabilities matter more. The hours you currently spend on routine email responses, meeting preparation, and document formatting are hours you're not spending on strategic thinking, relationship building, and creative work. Every task you successfully delegate to your assistant is time reclaimed for the work that actually differentiates you.

Here's the critical point: good delegation requires good management. You wouldn't hire a human assistant, hand them ambiguous instructions, and then blame them when the results are disappointing. Effective assistant management means clear communication, appropriate context, specific standards, and consistent feedback. Everything you're about to learn in this chapter—the SOP conversion, the context setting, the constraint engineering, the feedback loops—is good management practice applied to a new kind of assistant.

You are the executive. AI is your staff. The protocols that follow teach you to be a better manager of that staff.

WHAT WOULD YOU DO WITH TEN EXTRA HOURS?

Answer honestly: What would you do with that time?

The average professional who applies these delegation protocols reclaims seven to fifteen hours weekly—time currently consumed by routine work that could be handled by AI.

If your answer is "more of the same"—more emails, more meetings, more routine tasks—then delegation won't transform your career. It will just make you average faster. You'll process more volume at the same professional level, working harder without working strategically.

But if your answer involves something you've been postponing—the strategic project that never gets started, the relationships you've neglected, the thinking time you can never find, the skill development you keep deferring—then delegation becomes transformation.

The protocols in this chapter are tools. The transformation happens in how you reinvest the time they create. Know your answer before you build the tools.

THE "RELEASE" PROTOCOL

Your audit data tells a story. It reveals which tasks appear daily, which consume disproportionate time relative to their value, which follow predictable patterns you've executed hundreds of times before. Somewhere in that data are your highest-ROI delegation candidates—the tasks where AI assistance will produce the most significant return on the time invested in setting up that assistance.

Why Your Experience Makes You the Right Person for This

The professionals who delegate most effectively aren't AI experts—they're subject matter experts who know what quality looks like in their specific domain.

When AI produces a meeting summary, you can immediately assess whether it captured the right points because you understand meeting dynamics, organizational politics, and what details actually matter. When AI drafts an email, you can evaluate whether the tone is appropriate because you've written thousands of professional emails and internalized what works. When AI suggests a structure, you can judge whether it makes sense because you understand how your audience thinks.

This expertise doesn't come from studying AI. It comes from years of professional experience. Every email you've written, every meeting you've attended, every document you've produced has built pattern recognition that AI lacks. You're the quality control mechanism that makes delegation safe.

The professionals most skeptical of AI—those with deep expertise who've invested years building their craft—are precisely the ones best positioned to use it effectively. They have the judgment to verify quality. They know their standards intimately. They can recognize when output falls short.

AI works best when guided by expertise. Your experience isn't threatened by delegation—it's the asset that makes delegation work.

Identifying Delegation Candidates

The identification process begins with a single criterion: look for tasks that combine high repetition with low cognitive uniqueness. The activities you do frequently but that don't require your specific judgment to complete adequately. The weekly status report that follows the same structure every time. The meeting confirmation emails that vary only in names and dates. The document formatting that applies consistent standards to different content.

High repetition matters because setup costs are distributed across many uses. Teaching AI to handle a task you perform annually isn't worth the effort. Teaching AI to handle a task you perform daily pays dividends immediately.

Low cognitive uniqueness matters because it indicates the task doesn't require what makes you distinctively valuable. If a task requires your industry-specific expertise, your relationship history with a client, your judgment about ambiguous situations—that's not a Release candidate. If a task requires attention and time but could theoretically be performed by any competent professional with clear instructions, that's exactly what AI provides: a competent assistant following clear instructions.

Review your audit with these criteria. Mark the top five candidates —the tasks that appear most frequently and require the least of what's uniquely yours. These become your initial delegation targets.

The Sacred Task List

Before you delegate anything, write down three to five tasks you will never delegate—no matter how efficient AI becomes. These are your Sacred Tasks: work where your personal involvement is non-negotiable because the value comes specifically from you doing it.

Your Sacred Task list might include: handwritten thank-you notes to key clients, one-on-one mentoring conversations with direct reports, strategic decisions that define your team's direction, creative work that carries your authentic voice.

Defining what you won't delegate clarifies what you will. It prevents the slippery slope anxiety—the fear that delegation has no natural stopping point. When you know your boundaries, you can pursue efficiency within them without worrying that you're on a path toward professional irrelevance.

Review your Sacred Task list quarterly. Some tasks that seem irreplaceable today may become delegation candidates as your judgment evolves. Others will remain sacred for your entire career. The list isn't static, but it should always exist.

The SOP Conversion

You cannot delegate what you haven't defined. This principle applies to human assistants and AI assistants equally. Before AI can perform a task reliably, you need to understand that task explicitly enough to explain it.

Most professionals have never examined their routine work at this level. They execute tasks through accumulated habits—patterns so familiar they've become invisible. The email response happens. The meeting notes get cleaned up. The document gets formatted. But the specific steps, the decision points, the quality criteria—these remain implicit, stored in muscle memory rather than articulated knowledge.

AI delegation requires making the implicit explicit. You're converting tacit knowledge into a Standard Operating Procedure—a set of instructions clear enough that someone unfamiliar with you could follow them and produce acceptable results.

The Most Common Delegation Failure

The most common delegation failure happens before you even interact with AI: failing to define what "done" looks like.

Professionals routinely say "draft this email" or "summarize this document" without specifying what makes a good email or a good summary in their specific context. Then they're disappointed when the output doesn't match their unstated expectations.

The mistake feels efficient—why waste time explaining what seems obvious? But "obvious" only exists in your head. You have context the AI lacks. You have preferences you've never articulated. You have standards you've never written down.

Before delegating, answer: What would I look for when reviewing this output? What would make me say "yes, this works"? What would make me say "no, try again"? If you can't articulate these criteria, you're not ready to delegate the task—not because AI can't handle it, but because you haven't defined success clearly enough to recognize it.

The Four Conversion Questions

The conversion process involves four questions for each task you're preparing to delegate:

What is the input? Identify specifically what you're starting with. An email that needs a response. Raw meeting notes that need organization. A document that needs formatting according to brand standards.

What is the desired output? Describe specifically what you want to end with. The characteristics of a good result. The format, length, tone, structure that would meet your standards.

What are the steps between input and output? Break down the transformation. First, identify the key points. Then, organize them by theme. Then, apply consistent formatting. Then, add standard headers.

What are the decision rules? Identify the judgments involved. If the request is routine, respond with this template. If the request is urgent, flag for human attention. If the information is missing, ask this follow-up question.

This conversion exercise is valuable beyond AI delegation. It reveals inefficiencies in your current processes. It identifies tasks that are more complex than they seemed. It sometimes reveals that a task you've been doing for years could be simplified significantly.

Email and Communication Protocols

The average knowledge worker receives 121 emails daily and spends 2.6 hours managing them. Routine correspondence—the predictable exchanges that follow established patterns—consumes a significant portion of this time.

Not all email belongs in the Release category. Messages requiring nuanced judgment, relationship-sensitive communications, anything where tone matters as much as content—these remain in the Protect zone. But pattern-based correspondence offers substantial delegation opportunities.

Acknowledgment emails that confirm receipt without adding content. Meeting scheduling exchanges that coordinate times and locations. Status update requests that get the same response format every time. Information requests that can be answered by consulting existing documents.

For each communication type you're delegating, build a prompt that includes your standard response structure, your typical tone markers, and the specific information that varies per instance. The AI produces a draft. You review for accuracy and appropriateness—a thirty-second check rather than a five-minute composition.

The danger to avoid: sounding robotic. AI-generated correspondence often has a particular smoothness that feels impersonal. Counter this by injecting your voice markers after the draft is generated, or by training the AI with examples of your actual writing. The goal is correspondence that sounds like you, produced with less effort—not correspondence that sounds like everyone.

Administrative and Logistical Tasks

Meeting notes exemplify administrative work ripe for delegation. The raw transcript or rough notes exist. The organized, actionable summary needs to be produced. The transformation is largely mechanical—identifying key decisions, extracting action items, structuring information by topic or participant.

AI handles this transformation well because the task is primarily organizational rather than judgmental. The quality criteria are relatively objective: Are the key decisions captured? Are action items clearly identified? Is the structure logical?

Document formatting follows similar logic. Brand standards specify fonts, margins, header styles, color schemes. Applying these standards consistently requires attention but not expertise. AI can receive an unformatted document and return a properly formatted version—or at least get close enough that final adjustments take minutes rather than hours.

File organization, calendar coordination, research gathering, data entry from one format to another—each administrative task follows the same assessment: Is the transformation pattern-based? Are the quality criteria definable? Can you describe what "good" looks like specifically enough that a non-expert could recognize it?

Where the answers are yes, you've found Release candidates.

THE DELEGATION DIALOGUE

Effective AI delegation requires effective AI communication. The quality of output depends directly on the quality of input—the clarity of your instructions, the completeness of your context, the precision of your constraints.

Context Setting: The Onboarding Concept

When you hire a human assistant, you don't hand them tasks on day one without any background. You explain your role, your preferences, your standards, the context in which you operate. This onboarding period enables effective performance because the assistant understands not just what to do but why it matters and what constraints apply.

AI requires the same onboarding, compressed into each interaction.

Before asking AI to draft an email response, tell it who you are professionally, who you're writing to, what your relationship is, what outcome you're seeking. Before asking AI to summarize a document, tell it what you'll use the summary for, what level of detail matters, what aspects are most relevant to your purpose.

This context investment feels inefficient initially. Why spend thirty seconds explaining yourself when you could just ask the question? Because those thirty seconds dramatically improve output quality. Without context, AI defaults to generic assumptions. With context, it adapts to your specific situation.

The efficiency comes from building context templates. Your professional identity. Your typical communication style. Your standard quality criteria. Capture these in a document you can paste into new conversations, eliminating the need to reconstruct context from scratch each time.

Constraint Engineering

What you tell AI not to do often matters as much as what you tell it to do.

Unconstrained, AI tends toward the verbose, the formal, the hedge-filled. It produces output that covers all bases at the expense of directness. It includes caveats and qualifications that serve no purpose except liability avoidance.

Constraints shape this tendency into something useful.

Specify length: "Keep this under 100 words." Specify tone: "Write in a conversational register, avoiding formal business language." Specify structure: "Use three short paragraphs, not bullet points." Specify exclusions: "Do not use the phrase 'I hope this email finds you well' or any similar generic openings."

Each constraint removes degrees of freedom that would otherwise produce unwanted variation. The more precisely you constrain, the more reliably you get output that matches your standards.

This constraint discipline has a secondary benefit: it forces you to clarify your own preferences. When you tell AI to avoid certain phrases, you're articulating standards that were previously implicit. When you specify tone, you're defining your voice in explicit terms.

The discipline of constraint engineering makes you more aware of what you actually want.

The Feedback Loop

When AI produces inadequate output, the temptation is to abandon the attempt. The draft isn't right. Easier to write it yourself. Move on.

This response wastes the learning opportunity.

AI output that misses the mark tells you something about the gap between your instructions and your intentions. Either the instructions were unclear, or you're asking for something the AI struggles to provide. Either way, the failure is information.

Instead of rewriting the output yourself, tell the AI what was wrong and ask it to try again. "This is too formal. Rewrite in a more conversational tone." "You missed the main point, which is [X]. Reorganize to emphasize that." "The opening is generic. Start with a specific reference to our previous conversation."

This iterative refinement accomplishes two things. It often produces better output than your original prompt would have. And it teaches you what instructions produce what results, building your prompting skill through direct feedback.

Building Your Prompt Library

Effective prompts are assets. Once you've developed instructions that reliably produce good output for a specific task type, save them.

A prompt library eliminates repeated problem-solving. The first time you draft a meeting summary prompt might take ten minutes of iteration. The hundredth time takes ten seconds—copy the prompt, insert the specific meeting details, generate.

Organize your library by task type. Communication prompts. Analysis prompts. Formatting prompts. Administrative prompts. Within each category, capture the prompt text, notes on what works well, and any variations for different sub-cases.

Over time, this library becomes your personalized AI instruction set

—a collection of tested approaches that produce reliable results in your specific professional context. The compound value is substantial. Hours invested in developing good prompts are distributed across hundreds of future uses.

QUALITY CONTROL AND TRUST

Delegation without verification isn't delegation. It's abdication. The Human-in-the-Loop standard from Chapter 2 applies to every Release task: nothing AI-generated leaves your desk without human review.

The Spot Check System

You don't need to scrutinize every word of every AI output. You need to scrutinize enough to catch problems before they become consequences.

Implement a sampling method proportional to stakes and confidence. For high-stakes outputs—external communications, documents with your name attached—review thoroughly. For low-stakes outputs where you've established AI reliability—internal notes, draft organization—spot check a representative sample.

The ratio shifts over time. Early in your delegation practice, review everything carefully. As you build confidence in specific task types, reduce scrutiny for those while maintaining vigilance for new applications.

Random sampling matters more than predictable sampling. If you always review the first paragraph, you'll catch first-paragraph problems but miss errors buried deeper. If you always check formatting, you'll catch formatting issues but miss content problems. Vary your verification focus to catch the full range of potential failures.

Hallucination Hunting

AI confidently asserts things that aren't true. This characteristic, called hallucination, is fundamental to how large language models work—

they generate plausible text, not verified truth. Plausibility and accuracy overlap significantly but not completely.

Learn the red flags. Specific numbers without clear sources. Precise dates or statistics that you didn't provide. Citations to articles or studies —AI frequently fabricates these. Names of people, places, or organizations that you can't independently verify.

Any factual claim in AI output deserves verification against a source you trust. This doesn't mean paranoid checking of every statement—context and common sense apply. But claims that carry consequences if wrong require confirmation before you pass them along.

The verification habit protects more than accuracy. It maintains your professional judgment. Professionals who stop verifying AI output gradually outsource their critical thinking. Professionals who verify consistently maintain the skeptical evaluation skills that distinguish expertise from information relay.

The 80% Rule

Perfectionism destroys delegation value.

If you spend as much time revising AI output as you would have spent creating from scratch, you haven't gained anything. The goal isn't perfect AI output. The goal is output that's good enough to serve as a foundation, with your final polish adding the remaining quality.

The Pareto Principle provides a useful frame: roughly 80% of value comes from 20% of effort. AI typically provides a solid 70-80% of acceptable output. Your contribution is the 20-30% that elevates adequate to excellent.

This doesn't mean accepting mediocrity. It means recognizing where marginal improvement requires disproportionate effort. The difference between 80% quality and 85% quality might require more work than the difference between 0% and 80%. Your judgment determines which improvements matter enough to justify that work.

Practically: set a time limit for AI output review. If the output meets the threshold within that limit, publish it. If it requires fundamental rework, abandon the AI approach for this instance and try a different prompt next time. Don't spend thirty minutes polishing a draft that should have taken AI ten seconds to get right.

"But if I delegate this work, will I lose touch with what my team actually does?"

The fear of losing touch is grounded in real management wisdom. Leaders who become too removed from operational reality make poor decisions. They lose credibility with their teams. They miss warning signs that would be obvious to someone closer to the work.

But there's a crucial distinction between losing touch and strategic distance.

Losing touch happens when you delegate and forget—when you hand off work and never think about it again, trusting outputs you haven't verified against standards you haven't defined. That's not what this chapter teaches. The protocols here require you to define quality criteria explicitly, review outputs against those criteria, and maintain verification systems that catch problems before they cascade.

Strategic distance is different. It's the appropriate separation between understanding how work should be done and personally executing every instance of that work. Every effective leader operates at strategic distance from some portion of their organization's activity. CEOs don't personally review every customer service email, but they understand customer service standards and hold teams accountable to them.

The work you're delegating through AI doesn't require you to be closer to it. It requires you to have defined what "good" looks like clearly enough that you can verify outputs against that standard. That's not losing touch—that's mature professional capability.

Stay in touch with standards, outcomes, and quality. Release the mechanical execution that your judgment no longer needs to personally oversee.

BUILDING RELIABILITY OVER TIME

Trust develops through accumulated evidence. As you delegate the same task type repeatedly and consistently receive acceptable output, your confidence in AI reliability for that specific application grows.

Track this development deliberately. Note which task types produce reliable results and which require frequent intervention. Note which prompt structures work consistently and which produce variable quality.

Over time, you'll identify a core set of tasks where AI assistance is genuinely trustworthy—where spot checking replaces detailed review, where your confidence is earned through repetition. You'll also identify tasks where AI consistently underperforms—where the effort of delegation exceeds the benefit, where human creation remains more efficient.

This differentiation is the goal. Not universal AI delegation. Calibrated AI delegation, informed by evidence about where the tool actually helps in your specific practice.

REINVESTING THE DIVIDEND

Release produces a dividend: time and cognitive energy reclaimed from routine work. The value of this dividend depends entirely on where you reinvest it.

The Vacuum Effect

Cyril Northcote Parkinson articulated the principle in 1955: work expands to fill the time available for its completion. Save two hours through AI delegation, and those two hours will disappear—absorbed into longer versions of remaining tasks, filled with new routine work that didn't exist before, consumed by the general expansion of activity that accompanies available capacity.

This isn't a moral failure. It's human psychology and organizational dynamics operating as expected. Without intentional direction, reclaimed time vanishes.

The counter-measure is explicit allocation. Before you gain the time, decide where it goes. Not vaguely—specifically. Name the project, block the calendar, commit the hours.

The Weekly Delegation Tracker

Create a simple tracking system that measures your delegation progress and reclaimed time. Each week, document three things:

Tasks Delegated This Week: List every task you handed to AI, even partially. Note whether the output was acceptable immediately, acceptable after one revision, or required significant rework.

Time Reclaimed (Estimate): Calculate rough time savings. If a task normally takes thirty minutes and AI handled it in five minutes of your time (prompting and review), you reclaimed twenty-five minutes. Track cumulative reclaimed hours across the month.

Reinvestment Log: Document specifically where reclaimed time went. Strategic project work? Relationship building? Deep thinking? Rest? The goal isn't to fill every minute—it's to ensure reclaimed time goes somewhere intentional rather than getting absorbed into more routine work.

Review your tracker monthly. Patterns will emerge: which task types produce reliable results, which require more supervision, and whether you're actually reinvesting saved time or letting it disappear into busywork expansion.

Strategic Rest

One legitimate reallocation: breaks.

This sounds counterproductive in a book about professional performance. You've freed up time. Why waste it on doing nothing?

Because cognitive freshness is a performance resource. The average knowledge worker maintains focused attention for limited periods before degradation sets in. After that threshold, continued work produces diminishing returns. The professional who works eight hours of degraded attention often accomplishes less than the professional who works six hours of fresh attention.

AI delegation offers the opportunity to work fewer degraded hours —not by leaving early, but by inserting recovery between intense work blocks. The thirty-minute walk after releasing routine correspondence. The genuine lunch break made possible by automated meeting prep. The fifteen minutes of deliberate disengagement between analytical tasks.

These breaks aren't wasted time. They're maintenance of the cognitive resource that your highest-value work depends on.

Deep Work Blocks

Research from the University of California, Irvine found that workers take an average of twenty-three minutes to return to full focus after a distraction. Every routine task that interrupts strategic work carries this recovery cost—not just the minutes spent on the task itself, but the extended time to regain focus afterward.

The average knowledge worker experiences forty-seven interruptions per day. If even half of those involve context-switching to routine tasks, and each carries a twenty-three-minute recovery cost, you're losing hours of focused capability daily—not to the tasks themselves, but to the cognitive transitions around them.

Delegation isn't just about freeing minutes. It's about protecting extended periods of uninterrupted focus for work that requires sustained attention.

The more common reallocation is strategic work—the thinking, planning, creating, and relationship-building that defines professional differentiation.

What project have you been postponing because you never have uninterrupted time? What strategic question deserves sustained attention but keeps getting squeezed by urgent demands? What relationship needs investment you haven't been able to provide?

These are your reinvestment targets. Name them explicitly. Allocate the reclaimed hours to specific priorities, with calendar blocks that protect the time against encroachment.

The allocation doesn't need to be ambitious initially. Five hours reclaimed from routine work, redirected to one strategic project. That's enough to produce meaningful progress where nothing was happening before. Meaningful progress builds momentum. Momentum builds into transformation.

THE MEETING PREP TRANSFORMATION

Before delegation, meeting preparation looked like this: twenty minutes gathering documents, fifteen minutes skimming emails for context, ten minutes drafting discussion points, five minutes formatting an agenda. Total: fifty minutes, often rushed, sometimes skipped entirely when time ran short.

After delegation, the same preparation happens differently: a three-minute prompt provides meeting context and asks for key points, decisions needed, and suggested questions based on attached documents. Five minutes to review and adjust the output. Ten minutes to add personal notes on relationship dynamics and strategic positioning—the elements that require your judgment.

Total: eighteen minutes, executed consistently because the barrier is lower.

The transformation isn't just faster. It's better. Previously, time pressure meant skipping context review or showing up underprepared. Now, comprehensive preparation happens reliably because the routine portion is handled. You arrive at meetings with better information and more energy for the human dynamics that actually determine outcomes.

THE RECLAIMED WEEK

Consider how the math works across a broader view.

McKinsey research found that knowledge workers spend only 39% of their time on role-specific work—the tasks they were actually hired to do. The remaining 61% goes to communication overhead, searching for information, and administrative coordination. Similar studies from RescueTime show that the average professional spends 2.6 hours daily on email alone.

A professional spending fifteen hours weekly on Release-eligible tasks—meeting prep, routine correspondence, status updates, document formatting—who captures even half that time transforms their weekly capacity.

Before: Strategic planning squeezed into Friday afternoons when exhaustion has set in. Project risk assessment happens reactively when problems emerge. Stakeholder relationships maintained through rushed conversations between meetings.

After: Tuesday mornings blocked for strategic planning, approached with fresh attention. Weekly risk reviews conducted proactively, catching issues before they compound. Stakeholder conversations given the time they deserve, building trust that pays dividends when projects hit inevitable obstacles.

The deliverables aren't different. The quality of thinking behind them is transformed.

This is what 'Release' makes possible—not the same work faster, but different work entirely. The cognitive clutter that consumed your weeks, cleared to reveal the strategic capacity that was always there, waiting for the bandwidth to express itself.

THE COGNITIVE CLUTTER that consumed your weeks begins clearing the moment you implement your first Release protocol. You've learned to identify what requires your judgment and what merely requires competent execution—then you've learned to delegate the latter systematically. The hours you reclaim aren't just recovered time; they're cognitive capacity freed for the work that actually differentiates your professional value. Release clears the foundation. In the next chapter, you'll explore 'Expand' - how to transform that reclaimed capacity into creative output you couldn't have produced when routine work was drowning you.

5

EXPAND THE CREATIVE

Creative work fails at predictable points in the value chain. The blank document. The undefined brief. The project requires original thinking but offers no obvious entry point. You encounter these moments and respond with characteristic inefficiency: procrastination that wastes hours, avoidance that delays projects, finally forcing something adequate onto the page when deadline pressure becomes unbearable.

The bottleneck isn't creative capability—you possess far more creative capacity than your output suggests. The bottleneck is initiation friction, the cognitive load required to generate the first viable direction from infinite possibility.

You know this friction intimately. That moment when you sit down to write and nothing comes. When you need to design a presentation structure and every option seems simultaneously right and wrong. When you face a client proposal and can't find the angle that makes the value clear.

Traditional productivity advice addresses this friction with willpower: just start, get words on the page, embrace imperfection. But willpower doesn't reduce friction—it just powers through it at significant energy cost.

AI eliminates initiation friction entirely by generating starting points on demand, transforming creative work from pure generation to strategic selection. This is 'Expand', the second step of the Professional Renaissance Method.

As Peter Drucker observed: "Creativity is not a talent. It is a way of operating."

The statement is easily misread as dismissing natural creative ability. Drucker meant something different. Creativity, in professional contexts, emerges from systems and practices more than from inspiration and gifts.

The creative professional who waits for inspiration produces inconsistently. The creative professional who operates systematically—who has processes for generating ideas, evaluating options, and developing selected concepts—produces reliably.

AI tools don't replace creativity. They enable better operating systems for creative work. They systematize the generation that used to depend on lightning-strike inspiration. They make the creative process reliable rather than random.

When you adopt AI-augmented creative practices, you're not diminishing your creativity. You're improving your operating system. You're moving from waiting for inspiration to engineering conditions where good work can emerge predictably.

The fear that using AI for creative work somehow disqualifies you from being creative makes sense. Your professional identity might be built on being "the ideas person," "the creative one," "the one who comes up with the campaigns." If a machine produces the raw material, what's left of that identity?

Here's the reframe: since when was creativity about the generation of raw material?

Consider film directors. They don't operate cameras, design costumes, build sets, write music, or—in many cases—write the screenplay. They direct. They choose. They shape. They say yes to this take and no to that one. They assemble performances, visuals, and sound into something that didn't exist before.

No one suggests Spielberg isn't creative because he didn't personally sew the dinosaur costumes.

The costume shop is a tool. AI is a tool. Creativity isn't in the tool—it's in knowing what to ask for, recognizing when you've found it, and shaping it into something that serves your vision.

Your creative identity doesn't reside in the generation of raw material. It resides in the taste that knows gold from garbage. The judgment that selects the right option from fifty possibilities. The vision that sees what could exist before it does. The refinement that takes rough output and polishes it until it's unmistakably yours.

These are harder skills than generation. They're also more valuable.

What you're actually good at—the real source of your creative value—was never the typing. It was the thinking. The AI handles more of the typing. You do more of the thinking.

That's an upgrade, not a diminishment.

"But My Work Will Become Generic"

This objection contains real danger—and real misunderstanding.

The danger is real: unrefined AI output does sound generic. It regresses toward the mean. It produces smooth, acceptable, forgettable prose that could have come from anyone. If you use AI-generated content directly, your work will lose distinctiveness.

The misunderstanding: that's not how this works.

AI-generated content is not your output. It's your input. It's the clay, not the pot. It's the lumber, not the house. It's the rough draft, not the final version.

Your voice doesn't come from the initial generation of words. Your voice comes from what you choose to keep, what you cut, what you add, what you twist into unexpected directions.

Think about how your voice actually developed. You read other writers. You absorbed their rhythms, their word choices, their structural approaches. Then you synthesized, selected, and shaped until something distinctively yours emerged. You didn't write from vacuum —you wrote from influence, filtered through your specific perspective.

AI is another influence. A fast, prolific, infinitely patient influence that produces raw material for your filter.

The professionals who lose their voice to AI are the ones who skip the filtering. They generate and ship. Of course the work sounds generic—they didn't add themselves to it.

The professionals who amplify their voice use AI to produce more raw material than they could produce alone, then apply more of their filter to more material. The output is more distinctively theirs, not less, because they're operating at the level where their judgment—not their typing speed—determines quality.

Your voice isn't threatened by abundance. It's threatened by laziness. Don't be lazy.

BREAKING THE BLANK PAGE

Creative work stalls at predictable points. The blank document. The undefined brief. The project that requires original thinking but offers no obvious starting point. Professionals encounter these moments and respond in characteristic patterns: procrastination, avoidance, finally forcing something onto the page that feels adequate rather than excellent.

The problem isn't lack of capability. It's initiation friction—the cognitive load of generating the first viable direction from infinite possibility.

AI dissolves initiation friction by generating starting points on demand. Not finished work. Not the creative vision itself. Starting material—options to react against, directions to evaluate, possibilities to accept or reject or combine.

The Ignition Prompt

Creative paralysis has a predictable structure. You face a deliverable—a presentation, a proposal, a campaign concept, a strategic recommendation. You know the general territory. You have expertise in the domain. But the specific approach remains undefined, and each moment of staring at the blank document reinforces the sense that nothing adequate will emerge.

The ignition prompt bypasses this pattern by requesting multiple starting points simultaneously. Instead of generating one idea under pressure, you ask AI to generate ten possibilities without pressure on any individual option.

The prompt structure follows a consistent pattern: describe the context, specify the output type, request multiple distinct options.

For a presentation: "I'm developing a presentation on [topic] for [audience] with the goal of [outcome]. Generate ten different opening angles I could take—varied approaches, not variations on the same approach. Each should be two to three sentences summarizing the angle and why it might work for this audience."

For a proposal: "I'm writing a proposal for [project/service] to [client/context]. Give me eight different ways I could structure the argument—different emphases, different ordering, different core claims about value."

For content: "I need to write about [topic] in a way that [goal—informs, persuades, entertains]. Suggest twelve possible hooks or entry points, ranging from conventional to unexpected."

The output from these prompts won't contain your final answer. That's not the purpose. The purpose is optionality—moving from zero possibilities to many possibilities, shifting your creative work from generation to selection.

Selection is cognitively easier than generation. Evaluating ten options against your criteria requires judgment, but it doesn't require overcoming the initiation friction of the blank page. Once you've identified the promising directions among the generated options, your creative energy focuses on development rather than conception.

Overcoming Generative Blocks

Creative blocks take different forms. Sometimes the problem is initiation—getting started at all. Sometimes the problem is expansion—developing an initial idea into something richer. Sometimes the problem is direction—knowing that something is wrong but not seeing the alternative.

Each form responds to specific prompting strategies.

For expansion blocks: "I have this initial concept: [describe]. Give me ten ways to develop it further—extensions, complications, applications, variations that might make it stronger or more interesting."

For direction blocks: "I'm working on [project] and something isn't working. The current approach is [describe]. Suggest five completely different ways I could approach this, as if starting over with different assumptions."

For perspective blocks: "Critique this concept as if you were [skeptical audience member, competitor, devil's advocate]. What weaknesses would you identify? What objections would you raise?"

The prompts use specific techniques: requesting multiple outputs rather than single answers, asking for variety rather than similarity, framing requests to elicit genuinely different options.

Persona-Based Exploration

Creative thinking benefits from perspective diversity. The same problem looks different when examined through different professional lenses, different value systems, different aesthetic preferences.

AI can simulate perspectives on demand.

"Approach this problem as a behavioral economist. What would you emphasize? What would you measure?"

"Analyze this concept as a design critic. What's working? What's confusing or unclear?"

"Evaluate this strategy as if you were our most skeptical board member. What concerns would you raise?"

These persona prompts don't produce expert-level analysis from those perspectives. They produce approximations that can shift your own thinking. The behavioral economist framing might surface considerations you'd otherwise miss. The design critique might identify confusion you'd normalized. The skeptical board member might articulate objections you'll eventually face.

You're not outsourcing expertise. You're using AI to prompt your own perspective-taking, generating viewpoints to react against and integrate.

ITERATION VELOCITY

Creativity compounds through iteration. First drafts improve through revision. Initial concepts sharpen through development. The gap between adequate and excellent usually closes through cycles of creation, evaluation, and refinement.

Traditional creative work imposes iteration costs. Each cycle takes time. Each revision requires energy. The number of cycles you can complete constrains how much improvement is possible within project timelines.

AI compresses iteration cycles by handling the mechanical work of revision, freeing your attention for evaluative judgment about what's working.

Rapid Prototyping

Before committing significant resources to a creative direction, you want to test whether the direction has potential. A full execution takes hours or days. A rough prototype takes minutes.

AI enables rough prototyping for text-based creative work. You can produce an outline for a presentation and evaluate whether the structure works before building slides. You can generate a draft email sequence and assess whether the messaging logic holds before polishing individual messages. You can create a rough version of a report section and determine whether the approach serves the purpose before investing in detailed writing.

The rough outputs have obvious limitations. They lack your voice, your nuance, your specific expertise. But they reveal structural issues—whether the logic flows, whether the length is appropriate, whether key elements are missing. These structural questions are often clearer in rough form than in your own mental projection of what you intend to create.

The process: generate a rough version quickly, evaluate it with the question "does this direction work," iterate on the direction before investing in execution quality. AI handles generation. You handle direction-setting and evaluation.

Style and Format Transfers

The same content serves different purposes in different forms. The detailed analysis that works in a report fails as an executive summary. The conversational language appropriate for some audiences sounds wrong for others. The structure that works for one context needs adjustment for another.

AI handles format transfers efficiently. "Condense this three-page section into three paragraphs that capture the essential points." "Rewrite this formal proposal in more conversational language." "Convert this written explanation into an outline for a verbal presentation."

These transfers are mechanical—they don't require your creative judgment, only your verification that the transfer preserved what matters. You provide the source content and the target format. AI handles the transformation. Your time goes to checking the result rather than performing the transformation yourself.

Format flexibility expands creative optionality. Ideas that seem locked into one format can be tested in others. The blog post can become a presentation. The presentation can become a document. The document can become a summary. Each format reveals different aspects of whether the underlying content works.

Combinatorial Creation

Novel ideas often emerge from combining existing elements in unexpected ways. The strategy from one industry applied to different contexts. The structure from one creative format adapted for another. The unexpected connection between concepts that seemed unrelated.

AI can generate combinatorial suggestions on request. "What would happen if we applied [concept A] from [domain A] to the problem we're facing in [domain B]? Describe three possible connections."

"Combine the structural approach of [example 1] with the tone of [example 2]. What would that look like?"

"What surprising analogies might illuminate [problem we're solving]? Suggest five comparisons from unrelated fields."

These combinations aren't guaranteed to produce value. Most won't. But the cost of generating them is low enough that even occasional insights justify the exploration. You're expanding the territory of what you consider, increasing the probability that something genuinely novel emerges.

Combinatorial prompts work best when you provide concrete examples rather than abstract descriptions. Instead of asking for "creative combinations," specify the actual elements you want combined. The more concrete your inputs, the more useful the combinations AI generates.

Visual Ideation

For professionals whose work includes visual elements—presentations, marketing materials, concepts for designers to execute—AI image generation offers a new form of rough prototyping.

Describe the visual concept you're imagining. Generate rough versions to evaluate whether the direction works. Iterate through variations to clarify what you actually want before engaging with designers or committing to production.

This applies most directly to professionals who lack visual skills themselves. The marketing manager who sees campaign imagery in their mind but can't sketch it. The presenter who knows they want a particular visual approach but can't articulate it to designers. The entrepreneur developing a brand who has feelings about aesthetic direction but no vocabulary to express them.

AI-generated images aren't finished work. They're visual drafts—concrete enough to evaluate, concrete enough to share with collaborators, concrete enough to serve as reference for actual production.

CURATING THE CHAOS

Generating abundant options creates a new problem: evaluating which options deserve development. The blank page paralysis gets replaced by abundance paralysis—too many directions, insufficient clarity about which direction serves the purpose.

This is where your expertise becomes essential. AI generates. You curate.

Selection Is the New Generation

The shift happening in professional creative work: the scarce skill is changing.

For decades, generation was scarce. Producing options required time and energy. Professionals who could generate quickly held advantage. Writer's block, ideation paralysis, and blank-page terror were genuine obstacles because the generation bottleneck was real.

Now generation is abundant. AI produces options at volumes no human could match. The generation bottleneck has been removed—or at least radically widened.

What becomes scarce in abundance? Selection. Curation. The ability to identify excellence among options. The judgment that separates merely adequate from genuinely excellent.

This is not a lesser skill. It's a harder skill.

Generation under scarcity required producing something viable under pressure. Selection under abundance requires recognizing excellence among noise—and noise, by definition, sounds plausible. Mediocre options are often articulate, well-structured, and defensible. They don't announce themselves as mediocre. They require trained eyes to detect.

Your professional development question is no longer "how do I generate more?" It's "how do I select better?" Different question. Different training. Different competitive advantage.

The Evaluative Judgment

Generating ten options is easy. Identifying which three deserve development requires judgment that AI doesn't possess—judgment rooted in your understanding of audience, context, strategy, and quality.

The evaluative process involves applying criteria systematically across generated options. Different projects require different criteria, but common dimensions include:

Relevance: Does this option actually address the problem or purpose? Surprisingly often, generated options drift toward tangentially related ideas that don't serve the stated goal.

Distinctiveness: Is this option genuinely different from others under consideration, or is it a variation on the same underlying approach?

Feasibility: Can this option be executed within the constraints of time, resources, and capability?

Strategic fit: Does this option align with broader goals, brand positioning, or organizational priorities?

Quality potential: Does this option have room to become excellent with development, or is it inherently limited?

Apply these criteria quickly. You're not developing the options yet—you're filtering. Quick passes eliminate obviously wrong directions. Second passes compare remaining options against each other. The goal is a short list worth investing development time into.

REFINEMENT AND PERSONALIZATION

AI output sounds like AI output. There's a characteristic smoothness—grammatically correct, topically relevant, but lacking the texture that makes writing distinctive. Direct adoption of AI-generated content produces work that feels generic.

The refinement step adds what AI cannot provide: your voice, your specific expertise, your personal relationship to the material.

Practical Refinement Techniques

Add concrete specifics. AI generates in generalities. Replace generic examples with specific ones from your experience or knowledge. Replace abstract claims with precise data. Replace vague language with exact description.

Adjust rhythm and syntax. AI sentences tend toward similar lengths and structures. Vary them. Add short punches. Break patterns. Interrupt smooth flow with deliberate roughness where emphasis helps.

Insert perspective markers. Your opinions, your emphases, your interpretation of what matters most. AI presents balanced summaries. You take positions.

Include relationship signals. References to shared history, acknowledgment of context, language that demonstrates you know the specific audience.

The refinement process takes time—but it's refinement time, not creation time. The structural work is complete. The direction is established. Your effort focuses on elevation rather than construction.

Voice Calibration

For professionals who want AI drafts that start closer to their natural voice, calibration involves training AI with examples of your actual work.

The method: Provide AI with several examples of your writing—

ideally from contexts similar to what you're creating. Ask it to analyze patterns in your style. Then, for subsequent drafts, instruct AI to match the patterns it identified.

"Analyze these three email examples and describe my writing style —sentence length, vocabulary choices, typical structure, tone markers."

Then: "Draft a response to this request in the style you just analyzed."

Calibrated output isn't perfect. But it's closer. The refinement effort decreases because the starting point is already shaped toward your patterns.

Maintaining Originality

The risk of AI-augmented creativity is homogenization. If everyone uses similar tools with similar prompts, outputs converge toward similar results. The efficiency gain comes at the cost of distinctiveness.

Protection against homogenization requires conscious attention.

Ensure the core insight remains yours. AI can help with structure, variation, and expression. But the central idea—the strategic insight, the creative spark, the positioning that defines value—should emerge from your thinking, not from AI suggestions. Use AI to explore and express your ideas, not to substitute for having ideas.

Introduce deliberate constraints that create distinctiveness. What would make this work unmistakably yours? What approaches would you never take, clarifying what you would? What characteristic patterns in your best work should appear here?

Review with the question: could anyone have produced this? If yes, something is missing. Distinctive work carries the imprint of specific perspective. Generic work could have come from anywhere.

CREATIVE STANDARDS

The shift from scarcity to abundance requires new quality frameworks. When generating options was expensive, any viable option had value.

When generating options is cheap, the standard rises—adequate is abundant, so only excellent justifies development.

The Blandness Danger

AI defaults toward the safe, the balanced, the average. It produces text that avoids strong claims, unusual formulations, or anything that might be wrong or offensive. The result is smooth, acceptable, and forgettable.

Professional distinction requires edges. The sharp opinion that some will disagree with. The unusual phrasing that sticks in memory. The perspective that could only come from someone who has thought deeply about the domain.

AI-generated content, used directly, flattens these edges. The tendency toward blandness compounds across an organization—as more professionals use similar tools with similar prompts, the collective output regresses toward the mean.

Counter this tendency deliberately. When reviewing AI drafts, ask: what would make this bolder? Where is this playing it safe when conviction would be stronger? What is this omitting to avoid disagreement?

When reviewing any AI-generated creative work, run it through three questions before considering it finished:

What would make this bolder? AI defaults to safe positions. Find the places where the draft hedges when conviction would serve the work better. Strengthen the claims. Remove the qualifications that dilute impact.

Where is this too smooth? AI produces grammatically correct, rhythmically predictable prose. Introduce texture—short sentences, fragments, deliberate violations of standard structure. Break the pattern that makes text feel machine-generated.

What's missing that only I could add? Look for places to insert specific experience, concrete examples from your actual work, details that demonstrate you've lived with these problems rather than just read about them. The human fingerprint can't be prompted—it has to be contributed.

Three questions. Thirty seconds. The difference between output that sounds like everyone else's and work that carries your signature.

The Humanity Requirement

AI cannot inject genuine human experience into creative work. It can describe emotions. It can reference typical human situations. But it cannot draw on actual lived experience, actual relationships, actual consequences of real decisions.

This is your advantage.

Where AI-generated content feels authentic comes from the human elements you contribute: the specific example from your actual professional life, the admission of uncertainty that comes from real stakes, the perspective shaped by actual consequences of previous choices.

These elements can't be prompted. They can only be contributed. AI produces the structure; you provide the substance that makes readers recognize shared humanity rather than processed information.

The integration point: after AI generates draft content, ask yourself what real experience you can add. Not fabricated examples, but actual encounters with the concepts at play. Not theoretical possibilities, but actual observations from work you've done. The human elements don't need to dominate—a few well-placed specifics anchor entire sections.

The Hundred-Option Discipline

Volume unlocks quality, but volume requires intention. Generating ten options feels natural. Generating one hundred feels excessive—but the best option among one hundred is typically better than the best among ten.

Practice deliberate over-generation for significant creative challenges.

For headline options: generate one hundred, not ten. Evaluate quickly, discard eighty, develop twenty, select three for testing.

For structural approaches: generate twenty outlines, not three. Compare them systematically, identify patterns in what's working, synthesize the best elements.

For opening hooks: generate fifty variations before committing to one. The forty-ninth might be the one that actually works.

This discipline feels inefficient initially. But generation is cheap. Your time investment goes to evaluation, not creation. And the quality improvement from selecting among abundant options typically justifies the evaluation time.

CREATIVE APPLICATION IN PRACTICE

Leo led marketing for a tourism company. Before adopting AI tools, his process looked like this:

Week one: Initial team brainstorm. Four hours, conference room, whiteboard, coffee. The team generated maybe twenty campaign concepts. Most weren't viable—they violated budget constraints, didn't match channel strategy, or overlapped with previous campaigns.

Week two: Leo refined the surviving five or six concepts individually. Each one got a half-day of development: headline options, basic messaging, rough audience targeting. By Friday, three concepts were presentation-ready.

Week three: Stakeholder review. Two concepts survived. One became the campaign.

Elapsed time to campaign selection: three weeks. Total human hours: approximately forty across the team.

After integrating AI tools systematically, Leo's process changed:

Day one, morning: Leo spent ninety minutes with AI generating campaign concepts. Not twenty concepts—over one hundred directional starting points. The AI produced variations he'd never have reached in a team brainstorm: unexpected angles, surprising audience framings, bold positioning statements mixed with conservative approaches.

Day one, afternoon: Quick filtering. Leo evaluated the hundred concepts against strategic criteria: brand fit, budget feasibility, channel appropriateness. Sixty didn't make the cut. Forty went into a "develop further" list. Ten went into a "promising" shortlist.

Day two: Development. Leo used AI to rough out the ten promising concepts—not polished, but enough to evaluate whether they had legs. By the end of the day, three concepts were clearly superior, seven were eliminated.

Day three: Presentation preparation. Leo refined the three concepts to stakeholder-ready quality, adding the strategic rationale and human insight that AI couldn't provide.

Elapsed time to campaign selection: three days. Total human hours: approximately fifteen (Leo only—no team coordination overhead).

But the real improvement wasn't efficiency. It was quality.

The campaign Leo presented came from territory the team brainstorm would never have reached. The angle was genuinely unexpected—not because it was random, but because the broader exploration surfaced a strategic opportunity the constrained brainstorm would have missed.

Leo's creative value increased precisely because he stopped spending that value on generation mechanics and redirected it to selection judgment and strategic development.

The team didn't disappear. They shifted to execution—where their skills mattered for implementation—rather than ideation, where AI could search broader and faster. The work improved. The team's work improved. Only the bottleneck changed.

THE COMPOUND RETURN ON CREATIVE EXPLORATION

When you systematically explore more creative territory than you could explore alone, the benefits compound beyond any single project.

First-order benefit: better work on the current project. More options searched means better options are selected. Straightforward.

Second-order benefit: expanded creative vocabulary. The exploration exposes you to approaches you hadn't considered, phrasings you wouldn't have generated, structural possibilities outside your habitual patterns. These become available for future work even when you're generating without AI assistance.

Third-order benefit: calibrated taste. Evaluating fifty options develops evaluative muscle that evaluating five never could. You develop faster recognition of quality and faster rejection of mediocrity. Your internal filter sharpens through practice.

Fourth-order benefit: confidence in creative capacity. When you've consistently found excellent options through systematic exploration, the anxiety of "what if I can't find anything good" diminishes. You've proven to yourself that exploration works. Creative blocks become less terrifying because you have a reliable method for breaking them.

The hour you invest in AI-augmented exploration pays dividends across all future creative work. The skill compounds. The vocabulary expands. The taste is refined. The confidence builds.

This is what amplification means in practice: returns that multiply rather than add.

YOUR FIRST WEEK OF CREATIVE EXPANSION

The full chapter contains detailed tactics. But if you could only do a few things to transform your creative process this week, here they are.

Day 1-2: Adopt the Ignition Prompt

Choose one creative task you're currently facing. Before generating any ideas yourself, use this prompt structure:

"I'm creating [specific deliverable] for [specific audience] with the goal of [specific outcome]. Give me ten completely different approaches, ranging from conventional to unexpected."

Evaluate the ten options. Choose the one with the most potential. Develop it.

Success criteria: You started from AI-generated options instead of a blank page. You selected rather than generated your initial direction.

Day 3-4: Practice Volume Selection

For a new creative task, generate thirty options instead of ten. Use multiple prompts with different angles on the same problem.

Quick-filter to ten options. Then to five. Then choose one.

Success criteria: You evaluated abundance rather than scarcity. You practiced rapid selection among many possibilities.

Day 5: Add Your Voice

Take the AI-generated foundation from days 3-4. Deliberately refine it:

Add one specific example from your actual experience. Replace three generic phrases with language you'd actually use. Insert one opinion or perspective that only you would hold.

Success criteria: The final output sounds like you, not like AI. Someone who knows you would recognize your voice in the work.

Day 6-7: The Hundred-Option Challenge

Choose a task where you typically generate few options: headlines, subject lines, opening lines, or hook concepts.

Generate one hundred options. Don't evaluate during generation. Just collect.

Then filter ruthlessly. Mark the five that create the strongest reaction. Develop the best one.

Success criteria: You pushed past the point where the generation felt productive. You discovered something in the later options you wouldn't have found in the first ten.

After one week:

You've practiced the core skills of AI-augmented creative work: using AI to bypass initiation friction, selecting from abundance rather than developing from scarcity, adding human voice to AI-generated foundations, and pushing volume to discover unexpected quality.

These aren't one-time exercises. They're practices to repeat until they become automatic. The week establishes the patterns; the following months develop fluency.

Start simple. Start now.

WHAT COMES NEXT

This is 'Expand' applied to real professional practice. Not creativity outsourced, but creativity amplified. Not human judgment replaced, but human judgment applied to richer options than human generation alone could produce.

The 'Release' step cleared the routine work that was consuming your bandwidth. The 'Expand' step redirects that bandwidth toward creative amplification—more options, faster iteration, broader exploration of possibility space.

The next step shifts from generative work to analytical work. In the next chapter, you'll explore 'Partner' - how to collaborate with AI on research, synthesis, and the interpretation of complex information. The creative abundance of 'Expand' meets the analytical rigor of Partnership—different applications of the same fundamental principle that AI handles volume while you provide judgment.

6

PARTNER IN ANALYSIS

The average knowledge worker spends twenty-three hours per week reading and analyzing information—nearly 60% of their professional time—yet produces strategic insights during only four of those hours. The rest dissolves into mechanical processing: extracting key points from reports, cross-referencing sources, searching for patterns in data, condensing fifty pages into five useful paragraphs. You're intellectually capable of brilliant analysis, but you're computationally limited. Your working memory holds seven items. Your attention span fragments after ninety minutes. Your pattern recognition breaks down beyond twenty examples.

These aren't character flaws—they're biological constraints. Yet, analytical demands keep escalating. More data. More reports. More complexity. More decisions requiring synthesis across domains. This chapter introduces the third step of the Professional Renaissance Method: 'Partner'. Not replacing your analytical judgment, but removing the computational bottleneck that prevents you from applying it. AI handles tireless processing and perfect recall. You provide contextual wisdom and strategic interpretation. Together, you accomplish analysis that neither could achieve alone.

THE ANALYST ASSISTANT

You've hired a junior analyst. Impossibly junior—fresh from a program that taught them everything and nothing simultaneously. They've read every business book, every research methodology, every case study in existence. They remember all of it perfectly. They retrieve any information in seconds, cross-reference thousands of sources, and work around the clock without fatigue.

They also have zero judgment.

This junior analyst doesn't know which information matters and which is noise. They can't distinguish statistically significant patterns from meaningless correlation. They'll confidently present fabricated case studies alongside real ones because they can't tell the difference. They'll analyze your competitor's strategy without understanding that stated strategy differs from actual strategy. They'll summarize reports perfectly while missing the most important implication because it required reading between lines.

This is your AI analytical partner. Extraordinarily capable in ways no human junior analyst could match. Extraordinarily limited in ways no human junior analyst would be.

The metaphor clarifies your role. You're not checking work because you don't trust them—though skepticism is warranted. You're providing contextual judgment they fundamentally lack. You know which patterns in customer data actually predict behavior and which are data collection artifacts. You recognize when sources are authoritative versus merely sound authoritative. You understand the difference between what your organization says it values and what it actually rewards.

Consider how you'd manage an actual junior analyst with this profile. You wouldn't have them work independently on high-stakes analysis—that's disaster. But you wouldn't ignore their capabilities and do everything yourself—that wastes resources. You'd structure their work carefully. Give specific, bounded tasks. Review output before it reaches anyone else. Use their processing power for heavy lifting while you provide interpretation.

The same principles apply to AI partnership. Your junior analyst works for you. They enhance your capacity. They never replace your judgment.

CRITICAL WARNING

The most damaging analytical mistake isn't a wrong number or bad pattern—it's delegating the entire analytical arc rather than specific components.

Component delegation: "Summarize this report, focusing on operational changes and their stated rationale." You've bound the task. You've specified what matters. You maintain the analytical frame.

Arc delegation: "Analyze this report and tell me what I should think about it." You've surrendered the frame. You've asked AI to determine what matters. You've ceded interpretation.

Arc delegation fails because AI doesn't know your strategic context. It doesn't know what questions you're answering, what decisions you're informing, what constraints you're operating within. It will produce confident analysis entirely irrelevant to your actual needs. Worse, the confident presentation obscures the irrelevance.

Keep the analytical arc. Delegate specific tasks within it.

SUMMARIZATION AS STRATEGIC TOOL

Professionals drown in information while starving for insight. The research report, the meeting transcript, the industry analysis, the email chain—each contains signals buried in noise, relevant conclusions obscured by supporting detail. Traditional response: read everything (time-consuming) or skim selectively (risk-prone).

AI summarization changes the constraint. Mechanical extraction—identifying key points, condensing lengthy passages, organizing information hierarchically—happens instantly. Your analytical work begins where AI output ends: evaluating whether the summary captures what actually matters, identifying what was emphasized versus what should be emphasized, recognizing patterns across multiple summaries.

The structure:

First: Identify the analytical question you need answered. What specific insight are you seeking? What decision depends on understanding this material? What gaps in your current knowledge will this fill?

Second: Provide AI with source material and your analytical frame. "Summarize this quarterly report focusing specifically on operational changes and their stated rationale. I need to understand what strategic pivots they're making and why." The frame ensures the summary serves your actual need rather than providing generic condensation.

Third: Evaluate output against your expertise. Does this summary match your understanding of what's significant? What did AI miss that you recognize as important? Where did it emphasize details that don't matter to your question?

The time equation shifts dramatically. A fifty-page report might require ninety minutes to read thoroughly and twenty minutes to summarize in your own notes. With AI summarization: five minutes reviewing AI summary, ten minutes checking key sections the summary flagged, five additional minutes verifying understanding. Twenty minutes total with better coverage because AI processing doesn't suffer attention fatigue or confirmation bias.

Apply this systematically. Meeting transcripts condense to decision points and action items. Customer feedback aggregates into thematic patterns. Competitive analysis reveals strategic movements. Your analytical capacity expands not through faster reading but through faster processing of what's worth reading.

PATTERN RECOGNITION ACROSS DATA

Humans excel at recognizing patterns in small datasets. The marketing manager reviewing ten customer conversations spots common objections. The consultant analyzing three competitor strategies identifies shared vulnerabilities. The leader observing team dynamics notices recurring friction points.

Scale breaks this capability. Twenty customer conversations overwhelm working memory—you remember the most recent or most extreme, missing patterns that require holding all twenty simultaneously in view. Fifty competitor moves exceed attention span. One hundred team interactions create noise obscuring meaningful signal.

AI pattern recognition operates differently. Perfect recall. Simultaneous consideration of all data points. No attention fatigue. No recency bias. The capability to hold thousands of examples in computational view and identify statistical patterns humans cannot detect through observation alone.

The application requires you to define what patterns matter. AI finds correlations. You determine which correlations have causal or strategic significance.

For customer sentiment: "Analyze these two hundred customer support transcripts. Identify recurring themes in complaints, noting frequency and which themes correlate with account cancellation versus renewal."

For market analysis: "Review this dataset of competitor product announcements over eighteen months. What patterns emerge in their timing, positioning, or feature emphasis?"

For operational efficiency: "Examine these project completion records across teams. Where do delays cluster? What factors correlate with on-time delivery?"

The output provides statistical clarity manual analysis would miss. The customer complaints that seem uniformly distributed actually cluster around two specific issues. The competitor timing that appears random follows seasonal patterns tied to industry events. The project delays that feel unpredictable correlate strongly with specific handoff points in workflow.

Your analytical work involves interpreting these patterns. Correlation isn't causation. Statistical significance isn't strategic significance. The pattern AI identifies requires human judgment about whether it's actionable, whether it reflects underlying dynamics worth addressing, whether it's signal or coincidence.

SENTIMENT AND SUBTEXT

Written communication carries emotional content beyond literal meaning. The client email saying "fine" might signal satisfaction or resignation. The team member responding "I understand" might mean agreement or frustrated acceptance. The market reaction described as "measured" could indicate approval or skepticism.

Humans read subtext through experience—recognizing language patterns that signal emotional state. AI approximates this at scale, analyzing word choice, syntax, and phrasing patterns across large text volumes to identify emotional tone.

The application works best with volume. A single customer email analyzed for sentiment produces unreliable results. One hundred customer emails analyzed for sentiment distribution reveals whether your customer base skews positive, negative, or neutral—and which specific issues generate which emotional responses.

For stakeholder communication: "Analyze this year of team Slack messages in our project channel. What's the general sentiment trend? Where does sentiment drop?"

For market positioning: "Review social media mentions of our brand versus competitors. What emotional associations appear most frequently for each?"

For customer success: "Examine feedback from churned versus retained customers. What sentiment patterns differ between groups?"

The outputs require calibration against reality. AI sentiment analysis works better for some contexts than others. Professional language differs from social media language. Technical domains use vocabulary that sentiment models may misread. Your role involves verifying whether detected sentiment patterns match your qualitative understanding from direct interaction.

RESEARCH ACCELERATION

Professional research traditionally involves multiple steps: identifying relevant sources, locating them, reading for pertinent information, extracting key points, organizing findings, synthesizing insights. Each step consumes time. The total time from "I need to understand X" to "I understand X sufficiently to act" extends across hours or days.

AI compresses this timeline by handling information retrieval and preliminary extraction. The research question becomes: what specific knowledge gap am I filling? Then: what sources would contain this information? AI searches, retrieves, extracts initial findings, which you evaluate for accuracy and completeness.

This requires verification discipline. AI makes research mistakes—citing sources incorrectly, misrepresenting findings, inventing citations that sound plausible but don't exist. The acceleration comes with mandatory cost: you must check every significant claim, verify every important citation, validate every quantitative assertion.

The verification burden is less than the research burden. Reading five sources to confirm what AI extracted takes less time than reading fifty sources to extract information yourself. But verification is non-negotiable. Skip verification and you risk building strategy on hallucination.

The research acceleration enables higher-order work: asking better questions because you can iterate through research quickly, exploring adjacent domains because preliminary investigation cost drops, testing hypotheses rapidly because evidence gathering no longer bottlenecks analysis.

THE SOCRATIC METHOD

Rather than reading a document start to finish, interrogate it. Ask questions of the material. Probe for weaknesses. Request counter-perspectives. This transforms passive reading into active investigation.

The technique involves treating AI as an analytical intermediary between you and source material. You're not asking AI for its opinion—you're asking it to extract specific analytical perspectives from the material itself.

"What is the weakest argument in this strategic plan? Where does the logic break down?"

"This report claims X will lead to Y. What assumptions underlie that causal claim?"

"The proposal emphasizes benefit A but mentions cost B only briefly. What would a critic focus on?"

These prompts force analytical rigor. They surface objections you might not consider spontaneously. They reveal logical gaps you might overlook in favorable reading. They ensure you've examined material from multiple angles before accepting conclusions.

The outputs require judgment. AI might identify pseudo-weaknesses—points that seem vulnerable but aren't actually problematic when understood properly. Or it might miss genuine weaknesses because it lacks domain expertise to recognize them. Your evaluation separates valid criticism from analytical artifacts.

Apply this systematically to important documents. Before accepting strategic recommendations, interrogate them. Before endorsing analysis, probe it. Before committing resources based on research, challenge it. The Socratic approach catches flaws before they become costly mistakes.

GAP ANALYSIS

Comprehensive analysis requires identifying not just what's present but what's absent. The strategy that ignores competitive response. The proposal that omits implementation challenges. The research that neglects alternative explanations. Spotting gaps requires holding the full scope of relevant considerations while examining what the material addresses.

AI can systematically check for common analytical gaps. "What stakeholder perspectives does this proposal not address?" "What risk categories does this analysis omit?" "What implementation requirements does this plan leave unspecified?"

Gap identification works best when you provide explicit frameworks for completeness. Rather than asking AI to identify "any" gaps—too open-ended—specify what complete analysis would include, then ask what's missing.

For strategic planning: "Complete strategic analysis should address market dynamics, competitive positioning, operational requirements, financial projections, and risk mitigation. Which of these does this plan address inadequately?"

For decision support: "Sound decision analysis requires articulating decision criteria, evaluating options against criteria, addressing uncertainty, and considering implementation. What gaps exist in this analysis?"

For project proposals: "Comprehensive proposals specify scope, timeline, resource requirements, dependencies, risks, and success metrics. What's underspecified here?"

Gap analysis surfaces blind spots. Sometimes gaps are intentional —the analysis focuses narrowly by design. Sometimes they're oversights—important considerations lost in development. Your judgment determines which gaps matter enough to address before proceeding.

CROSS-REFERENCING SOURCES

Professional decisions often require synthesizing information from multiple sources. The internal data provides one perspective. The market research provides another. The competitive intelligence provides context. The historical precedent suggesting patterns.

Manually cross-referencing sources requires holding multiple information sets in working memory while identifying connections, contradictions, and complementary insights. Cognitive load limits how many sources you can effectively synthesize.

AI handles the mechanical work. "Compare these three market analyses. Where do they agree? Where do they contradict? What does each emphasize that others downplay?"

"This internal report says X. This industry study says Y. This competitive analysis says Z. What synthesis emerges across these three perspectives?"

"These historical case studies of similar strategies showed varied outcomes. What factors correlate with success versus failure across cases?"

Cross-referencing reveals patterns invisible in single sources. The market analyses agree on trend direction but diverge on magnitude—suggesting consensus on what's happening but uncertainty on how much it matters. The internal data contradicts external research on a specific point—flagging either measurement differences or genuine discrepancy requiring investigation.

Your role involves evaluating the synthesis. Are identified agreements meaningful or superficial? Do contradictions reflect different methodologies, different time periods, or genuine disagreement about reality? Which source deserves priority when they conflict?

SCENARIO MODELING

Strategic thinking requires considering multiple futures. The plan that succeeds if assumptions hold. The risks if assumptions fail. The opportunities if conditions shift unexpectedly. Scenario thinking traditionally happens through group discussion or individual imagination—limited by the number of plausible futures you can articulate and evaluate.

AI enables systematic scenario exploration. "We're planning to expand into market X based on assumptions A, B, and C. Model three scenarios: one where all assumptions hold, one where assumption A proves wrong, and one where external factor Y changes unexpectedly."

"For this product launch strategy, generate five distinct outcome scenarios ranging from best plausible case to worst plausible case, with explanations of what would need to happen for each."

"We're deciding between strategy Alpha and strategy Beta. What scenarios favor each? What conditions would make each the clear right choice?"

Scenario modeling doesn't predict the future. It surfaces contingencies to prepare for. The best case scenario suggests what success might require. The worst case reveals what risks need mitigation. The comparison scenarios clarify which strategy proves more robust across different possible futures.

Your strategic judgment determines which scenarios merit serious preparation. Some plausible scenarios matter so little that planning for them wastes resources. Others are sufficiently likely and consequential that they require explicit contingency planning. AI generates the scenario space. You decide which regions of that space deserve attention.

THE TRUST-BUT-VERIFY STANDARD

AI produces plausible output. Grammatically correct. Logically structured. Professionally formatted. This plausibility creates danger—the output looks trustworthy even when it's wrong.

Hallucinations compound this risk. AI confidently presents fabricated statistics, invented citations, and fictional case studies without signaling uncertainty. The claim sounds authoritative because AI expresses everything authoritatively.

The verification protocol: treat every factual claim from AI as provisional until confirmed through an independent source. Not skeptical of paralysis. Not trusting naivety. Provisional—it might be accurate, verify before depending on it.

For statistics: check the original source. Not just the statistic's magnitude but its definition, its measurement method, its recency. The 67% figure might be accurate but apply to a different population than you're addressing. The trend might be real but outdated by three years. The correlation might exist but not support the causal interpretation AI supplied.

For citations: verify they exist and say what AI claims they say. Academic sources prove particularly vulnerable to hallucination. AI might cite a paper that sounds perfectly relevant but doesn't exist. Or cite a real paper but misrepresent its findings.

For claims about current events: AI training data has a cutoff date. Information presented as current might be outdated. Market conditions might have changed. Regulations might have shifted. Company structures might have reorganized. Verify recency before building analysis on potentially stale information.

This verification discipline becomes automatic with practice. The time cost decreases as you develop efficient checking patterns. The risk reduction justifies the time—better to spend five minutes verifying than to discover misinformation after building strategy on it.

SOURCE TRACING

AI synthesis often combines information from multiple sources without clear attribution. This creates opacity about where specific claims originate. Verification requires source tracing—working backward from AI output to identify and check original sources.

The practical method: request explicit sourcing. "For each key claim in this analysis, provide the specific source: document title, author, publication date, and page number if applicable."

Or: "Analyze these five documents and synthesize key findings. For each finding, specify which document it came from using numbered references."

The explicit sourcing requirement forces AI to be more precise about origins. Output quality often improves simply because the sourcing constraint reduces vagueness and unsupported generalization.

When AI cannot provide sources for claims—because they emerged from synthesis rather than direct extraction—treat those claims as hypotheses to validate rather than facts to build on. Synthesis might produce valid insights, but without source grounding, you're accepting AI interpretation rather than verified information.

LOGIC VERIFICATION

AI produces text that follows grammatical logic without necessarily following reasoning logic. The conclusion might not follow from premises. The causal claim might lack supporting mechanisms. The recommendation might ignore contradictory evidence.

Logic verification requires examining analytical structure. Does the argument actually support the conclusion? Do the premises lead necessarily to the outcome? Are there logical gaps AI glossed over with confident language?

Common logical failures in AI analysis:

Correlation-causation confusion. AI identifies that X and Y move together and concludes X causes Y. The correlation might be real. The causation might be reversed, bidirectional, or absent entirely.

Incomplete comparison. AI evaluates option A and declares it superior without having actually evaluated competing options B and C systematically.

Unstated assumptions. The analysis depends on assumptions AI doesn't articulate. Without examining assumptions, you can't evaluate whether they're valid.

Cherry-picked evidence. AI emphasizes data supporting one conclusion while minimizing contradictory data. The imbalance creates a misleading impression of certainty.

Your verification catches these failures by asking: what would make this conclusion wrong? What evidence would contradict this? What assumptions would need to change for the opposite conclusion to be valid? If you can't articulate how the analysis could be wrong, you haven't understood it well enough to trust it.

REALITY CHECKING

AI lacks practical experience. It can describe business operations without understanding operational constraints. It can model financial projections without recognizing market realities. It can propose strategies without awareness of implementation challenges.

Reality checking requires applying your domain expertise to AI output. Does this analysis account for how this actually works? Does this recommendation recognize the constraints we actually face? Does this projection reflect what happens when you try to execute in real environments?

The consultant reviewing AI-generated competitive analysis spots immediately that the recommended positioning requires capabilities the organization lacks. The analyst examining AI financial projections recognizes assumptions that sound reasonable but contradict observed customer behavior. The operations leader reads AI process recommendations and identifies steps that would bottleneck in actual implementation.

This reality checking separates theoretical correctness from practical viability. AI might produce analysis that's logically sound but operationally impossible. Your expertise identifies where theory meets friction with reality.

THE BLIND SPOT PROBE

Before accepting any AI-assisted analysis as complete, ask yourself: what would change my interpretation of these findings?

This question forces you to articulate what your analysis has assumed rather than examined. The market analysis assumes stable macroeconomic conditions—what if conditions shift? The pattern recognition assumes data quality—what if collection methods were flawed? The competitive analysis assumes rational strategic behavior—what if the competitor is acting emotionally or politically rather than strategically?

The blind spot probe serves two functions. First, it identifies areas where additional verification or investigation might be warranted. Second, it calibrates your confidence in the analysis. Every analysis rests on assumptions. Knowing your assumptions doesn't make them true, but knowing them makes you appropriately humble about conclusions.

Make the blind spot probe habitual. Before any analysis leaves your desk, name what would change your mind.

FROM DATA TO MEANING

Analysis produces value through insight, not through information. Information describes what is. Insight explains why it matters and what to do about it. AI handles information aggregation efficiently. You provide the interpretive layer that converts information into actionable insight.

The translation from data to meaning requires contextual knowledge AI doesn't possess. You understand organizational priorities, stakeholder concerns, strategic constraints, competitive dynamics. You recognize which data points are genuinely significant versus merely interesting.

The synthesis process involves taking AI-generated analysis and adding strategic interpretation. The market research shows demographic shift X. Why does that matter for our positioning? What does it mean for product development? How should it influence resource allocation?

The competitive analysis reveals strategic move Y by key competitors. What does this reveal about their priorities? How does it change our competitive landscape? What should we do differently given this information?

The operational data shows pattern Z in customer behavior. What's driving this pattern? What does it predict about future demand? How should it shape our approach?

These interpretive questions turn analysis into strategy. The answers come from your expertise and judgment, informed by data AI processed but requiring human sense-making to become actionable.

DECISION SUPPORT ARCHITECTURE

Analysis serves decisions. The value of information lies in improving choice quality. This requires structuring analysis to support decision-making rather than just documenting findings.

Decision-quality analysis articulates: What decision requires making? What are the genuinely distinct options? What criteria matter for evaluation? How do options perform against criteria? What uncertainties affect outcomes? What should we decide?

AI handles the analytical legwork. You provide the decision architecture.

The structure: "We're deciding whether to launch product X. The real options are: launch now, launch after additional development, or don't launch. The key criteria are market readiness, competitive timing, resource availability, and risk exposure. Analyze available data to evaluate how each option performs against each criterion."

AI processes the analytical task. You evaluate the analysis against your strategic judgment. Does the market readiness assessment match your understanding from customer conversations? Does the competitive timing analysis account for moves you see coming? Does the resource assessment reflect actual capacity constraints?

The final synthesis combines AI analysis with human judgment to form clear recommendations. Not "here's what the data shows" but "based on this analysis plus strategic context, here's what we should do and why."

EXAMPLE

Julia's client faced a market entry decision with significant risk. The target market spanned three regions with different regulatory environments, competitive dynamics, and growth trajectories. Traditional analysis would require weeks—reading regulatory documents, analyzing competitor movements, interviewing customers, modeling financials. Synthesis meetings to integrate findings. Final report combining all perspectives.

She structured the AI partnership as a series of bounded tasks.

First: Summarize the regulatory environment in each region, focusing on barriers to entry and recent changes. This produced a comprehensive overview in twenty minutes that would have taken days of reading. She verified the summary against her knowledge of one region where she had previous experience—the AI had captured the dynamics accurately, building her confidence for the other regions.

Second: Analyze public competitor activities across all three regions over the past eighteen months. Identify patterns in expansion strategy, pricing moves, and market positioning. The pattern analysis surfaced something she wouldn't have caught manually—two competitors had quietly retreated from the second region, suggesting difficulties she should investigate.

Third: Cross-reference the market research reports she'd gathered against her client's specific capabilities. Where do their strengths align with market opportunities? Where do their weaknesses create risk? The cross-referencing synthesized perspectives across six different documents faster than she could have read any one of them thoroughly.

Her verification focused on high-stakes claims: regulatory barriers that would affect entry timeline, competitive dynamics that would shape positioning, market sizing that would influence investment scale. She found one error—a confused attribution of market share—and several gaps where the AI couldn't access information behind paywalls. But the foundation was solid.

The strategic evaluation that followed built on comprehensive processing she could not have completed in the available timeframe without AI partnership. Her recommendation to the client wasn't just faster—it was informed by broader context than traditional analysis would have permitted. The decision quality improved alongside the decision speed.

This is 'Partner' applied to professional practice. Not analysis outsourced, but analysis amplified. Not human judgment replaced, but human judgment applied where it matters most—in verification, interpretation, and strategic synthesis.

THE COMPOUNDING ADVANTAGE

The ultimate benefit of analytical partnership isn't speed or volume—it's decision quality. Consider what improves when you can process more information within decision timeframes. Recommendations rest on more comprehensive data rather than incomplete samples. Pattern analysis includes variables you'd have ignored under time pressure. Cross-referencing catches contradictions that would have slipped past exhausted attention. The decision you bring to leadership isn't just faster; it's more likely to be correct.

This quality improvement compounds. Better decisions today create conditions for better decisions tomorrow. The strategic recommendation that accounts for industry dynamics you actually analyzed rather than guessed at positions your organization more accurately. The resource allocation based on actual pattern analysis rather than intuition reduces costly mispositioning. Over time, the gap between your decision quality and that of professionals still constrained by unassisted analysis grows.

The benefit isn't efficiency—efficiency is the mechanism. The benefit is being right more often about things that matter.

7

THE TRUST PREMIUM

The market is dividing into two categories: automated commodity services and high-trust relationship services. The middle ground is compressing as AI handles more commodity work at lower cost. This creates opportunity if you position correctly. If your value depends primarily on executing defined tasks, AI steadily erodes your position. If your value centers on relationships, judgment, and trusted counsel, AI potentially strengthens your position by removing you from commodity comparison. The professionals who command premium pricing in five years will be those who built sufficient relationship capital that clients choose them for judgment-dependent work rather than capability-dependent work. This chapter shows you how to build that capital while everyone else optimizes it away.

You've spent three chapters learning to optimize. You've released routine tasks to AI. You've expanded creative output with machine partnership. You've accelerated analysis through AI collaboration. Your calendar has space it didn't have before. Your cognitive load is lighter. The optimization mindset is working.

And now I'm telling you to stop optimizing.

This isn't whiplash. It's boundaries.

Optimization works on processes with predictable inputs and measurable outputs. Manufacturing processes. Administrative workflows. Information processing. These are systems. Systems improve through efficiency gains. There's no soul in an email template that needs protecting. There's no relationship between you and your filing system that suffers when you automate it.

Human relationships aren't systems. They're organisms.

Organisms don't respond to efficiency interventions the way systems do. When you optimize a relationship—sending templated messages instead of personal ones, scheduling "check-in calls" on a recurring calendar like any other task, tracking touchpoints in a CRM like units on an assembly line—you're treating an organism like a system. The relationship starts dying.

The person on the other end can feel it. Maybe not consciously. But something shifts. Your efficient message lacks weight. Your systematic call feels hollow. Your optimized outreach lands wrong. The efficiency signals that you're managing them rather than connecting with them.

The time you saved needs different deployment, not continued optimization.

The saved hours from previous chapters are meant to fund the inefficient work of human connection. They create space for the two-hour coffee that could have been a fifteen-minute call. They allow the handwritten note that a template would have handled in seconds. They enable full presence in a meeting that would have been split attention otherwise.

You didn't optimize to have more optimization. You optimized to have more humanity.

The mindset shift: optimization is how you clear the runway. Connection is where you actually fly.

THE GARDEN AND THE ASSEMBLY LINE

Your professional network operates as two different spaces simultaneously.

The first is an assembly line. Tasks arrive, you process them, outputs emerge. Speed matters. Consistency matters. Each unit is identical to the last, and efficiency gains compound. Double your speed, double your output. Automate a step, eliminate it forever. The assembly line rewards optimization, standardization, and relentless improvement. This is where AI excels. This is where you should absolutely use it.

The second is a garden.

Gardens don't respond to optimization the way assembly lines do. You cannot rush a tomato. No efficiency technique will make the beans mature faster. Doubling your watering doesn't double your harvest—it drowns the plants. The garden operates on different logic: patient attention, seasonal timing, and care that can't be systematized.

Professional relationships are gardens, not assembly lines.

Consider what happens when you try to optimize a garden like a factory. You install automated irrigation to save time. Efficient, certainly. But automated irrigation doesn't notice that the corner bed drains differently and needs less water. It doesn't see the fungus starting on the lower leaves. It delivers precisely measured water on precisely scheduled intervals, regardless of whether yesterday was rainy or the soil is already saturated.

The garden suffers not from lack of resources but from lack of attention.

This is what happens when an efficiency mindset creeps into relationship maintenance. You automate birthday messages—efficient, consistent, never forgotten. But automated messages can't notice that someone's tone has been off lately, that they're posting less frequently, that their recent updates hint at professional struggle. The message arrives on schedule, perfectly formatted, completely hollow.

You are the gardener of your professional relationships. AI is a tool in your shed. Use it well. Never confuse the tool for the work.

THE HUMAN-ONLY ZONE

What happens when you automate relationship work:

You use AI to write thank-you notes. Draft congratulations emails. Compose condolences. Each one sounds fine. Professional. Appropriate. Completely empty.

People can tell. Maybe not consciously. But they sense it. The message hits wrong. The warmth doesn't land. Something feels off.

Anthropologist Robin Dunbar's research on human social networks reveals a layered structure. Humans maintain approximately five intimate relationships, fifteen close ones, fifty friendships, and 150 stable social connections. These aren't arbitrary preferences—they reflect cognitive limits on how many relationships the brain can actively track and maintain. No amount of effort, systems, or tools can expand the fifteen-person limit on deep relationship maintenance. You can extend reach through technology. You cannot extend depth.

Dunbar writes: "Friendships are the single most important thing affecting our health, wellbeing, and happiness. Yet maintaining friendships is the first thing we drop when life gets busy."

When AI promises "relationship management at scale," it's offering to expand something that can't be expanded. What it actually delivers is the appearance of relationship without the reality of it. A hundred automated touchpoints never equal one genuine conversation.

You need boundaries. Clear lines around what you'll never automate.

Start with emotional moments: delivering bad news, responding when someone shares good news, apologizing, thanking someone who really helped. These need you—not because writing them takes long, but because the person on the other end needs to know you spent real attention on them.

Add nuanced conversations: negotiations, performance discussions, offers, rejections. AI can't read what's not said. It can't catch the subtle signals that tell you whether to push or back off.

Include depth-building moments: mentoring, strategic advice, career guidance. These conversations build connection through attention quality. When someone senses you're half-listening while mentally drafting your next prompt, relationship development dies.

The decision rule: Before you automate any communication, ask—does this interaction build or maintain a relationship that matters to my career? If yes, handle it yourself.

THE THREE TIERS

You can't give everyone equal investment. Accepting this changes everything.

Tier One (15 people): Every single interaction is personal. No automated anything. Full protection. These are the relationships that directly shape your professional trajectory—your biggest clients, the executives who control major decisions, your most valuable mentors and strategic partners. Every email you send them, you write. Every call, you're present. Every meeting, full attention.

Tier Two (30-50 people): Important touchpoints are personal; routine updates can be efficient. The conversation about project challenges happens face-to-face. The status update gets automated. You're protecting what matters while accepting efficiency where it doesn't damage relationship quality.

Tier Three (everyone else): Efficiency is fine. Template responses. Automated scheduling. AI-drafted follow-ups. The relationship doesn't depend on perceived personal investment because you're not operating at that depth.

The Discomfort of Choosing

The tier system works. But it's hard because you're explicitly deciding that some people get less of you.

We're taught that everyone deserves equal treatment. That limiting investment in some connections is cold or calculating. The tier system flies directly in the face of this conditioning.

Here's what you have to accept: you cannot be everything to everyone. The math doesn't work. Pretending otherwise doesn't make you generous—it makes you chronically overextended and consistently disappointing.

The professional who refuses to tier is the professional who says yes to every coffee request and then cancels half of them. Who responds to everyone "when I get a chance" and gets to tier-one relationships last. Who feels guilty about everyone constantly because they've promised everyone the same level of attention they can only deliver to fifteen people.

Name the tiers. Accept the discomfort. Then actually show up for the relationships that matter most—which you can only do because you stopped pretending you could show up equally for everyone.

People move between tiers. Review this quarterly. Adjust as situations change.

DEEPENING CONNECTION

You've got those fifteen hours back. Most people waste them filling saved time with more execution.

Early career, execution matters—you prove capability through consistent delivery. Mid-career, the equation flips. Your value stops correlating with output volume. It starts correlating with relationship quality, strategic positioning, and trust networks that generate opportunities.

AI just gave you space to make that shift. Use it.

The move: Reinvest thirty percent of reclaimed time—roughly three to five hours weekly—into relationship development. Not networking in the grab-coffee-pick-your-brain sense. Real development of meaningful connections.

Schedule longer conversations. Most professional interactions run fifteen to thirty minutes. These work for information exchange. They fail at relationship building. Extended conversations—ninety minutes, two hours—create space for substantive discussion.

Be fully present. No phone. No laptop. No checking notifications. Total focus. This is increasingly rare, which means it creates disproportionate impact.

Move beyond transactions. Share knowledge that doesn't benefit your immediate work. Make introductions without expecting returns. Offer perspective on their strategic challenges without charging. These asymmetric contributions build trust that transcends transactions.

Listen more than you speak. Most people want to be understood more than they want to understand. The professional who creates space for others to think aloud becomes valuable regardless of specific expertise. AI cannot do this. It can provide conversation starters. It cannot listen with authentic curiosity.

AI AS RELATIONSHIP SUPPORT

The distinction matters. AI as relationship replacement damages trust. AI as relationship support enhances your capacity for authentic connection.

Think of AI as invisible infrastructure that makes human interaction better. Not the interaction itself—the preparation, follow-through, memory, and analysis that surrounds interaction.

Contact relationship memory: After each significant interaction, spend five minutes documenting key business points, personal details mentioned, commitments made, and current priorities. Before the next interaction, review. You enter conversations with complete context—you remember the daughter's college choice, the vendor frustration, the board pressure. This makes you seem remarkably attentive. You are attentive—you're using technology to supplement biological memory limits.

Pre-meeting research: Ten minutes with targeted prompts generates comprehensive briefing on new contacts. You walk in informed about their recent initiatives, aware of their public positions, equipped with relevant questions. This demonstrates respect.

Post-conversation follow-through: Meetings end with commitments. Most professionals handle these inconsistently. AI extracts every promise and drafts follow-ups. Commitments actually happen. Reliability reputation strengthens.

Relationship preparation: Before difficult conversations—performance feedback, negotiations, conflict resolution—use AI as a practice partner. Test different framings, identify potential reactions, prepare for likely objections. You enter actual conversation with less anxiety and clearer strategy.

The pattern: AI handles mechanical cognitive work surrounding relationships—memory, research, preparation, follow-through—so you can direct full attention toward relationship work itself.

THE TRUST ADVANTAGE

Professional markets increasingly divide into two categories: automated commodity services and high-trust relationship services. The middle ground compresses as AI handles more commodity work at lower cost.

This creates opportunities for professionals who position correctly.

Research on friendship formation suggests it takes roughly fifty hours of interaction to develop a casual friendship and over two hundred hours for a close one. Professional trust-building likely follows similar patterns. The client who comes to trust you as their strategic advisor—not just their vendor—becomes that way through accumulated hours of genuine engagement, not through efficient touchpoints.

Differentiation through authentic presence: Pre-AI, efficiency generated competitive advantage. AI-enabled market? Efficiency becomes table stakes. Everyone has AI accelerating responses. But authentic presence cannot be automated at scale. The professional who gives you two hours of undivided attention remains scarce. Scarcity creates value.

Access through relationship capital: Professional opportunities are distributed through networks, not markets. The introduction that leads to your next major client comes from someone who trusts you. AI accelerates capability in most domains. It cannot accelerate trust building. The relationship that generates your next opportunity five years from now requires five years of consistent engagement to develop.

Premium positioning: Markets pay premium prices for work requiring high trust and sophisticated judgment. Judgment-intensive work resists commoditization. This trust premium increases as AI handles more commodity work.

THE QUARTERLY REVIEW THAT SAVED A CAREER

David spent twelve years building expertise in enterprise sales. He was known as a closer—the rep clients requested when deals got complicated. His technical reputation was bulletproof.

What David hadn't noticed was that his network was atrophying.

He'd spent so many years being pulled into emergency deal cycles that relationship development had collapsed into relationship maintenance. Old mentors had retired without introductions to successors. Colleagues had moved companies without maintaining connection. Clients had closed contracts without ongoing touchpoints. He was technically exceptional and relationally impoverished.

The crisis came when his company restructured. David's vertical was being absorbed into a larger division. His position was being "evaluated." He needed advocates—people with political capital who would spend it on his behalf.

He looked at his network and realized: he had contacts, not relationships. People who knew his name, not people who would fight for him.

The recovery took eighteen months of deliberate relationship investment. First, the quarterly audit: Who actually matters? Who has influence? Who might advocate if he'd given them reason to? He identified fifteen people—barely—who qualified for tier-one treatment.

Then, the rebuild. Monthly substantial touchpoints with each. Not automated check-ins—real conversations where he listened more than spoke. He learned about their challenges. He offered help without expectation. He became a person who noticed them, not just someone who needed them.

When the next organizational shift came, David had advocates. People called on his behalf. Defended his value in rooms he wasn't in. Created opportunities that weren't publicly posted. The technical expertise that had always been there finally had the relationship infrastructure to leverage it.

David now dedicates three hours weekly to relationship development. The investment seems to violate every productivity principle he once followed—yet he's come to understand these conversations shape his career more than any deliverable ever could.

The quarterly audit that showed him an empty relationship portfolio probably saved his career. He caught the atrophy before it became irreversible.

WHAT ENDURES

Strip away the tactics, the tools, the specific techniques, and three principles remain that will survive any technological evolution:

First: attention is currency. The scarcest resource in any economy is human attention—genuine, focused, undivided presence. As everything else becomes more abundant through automation, attention becomes more valuable.

Second: trust compounds. Every authentic interaction deposits trust that generates returns for years. Every hollow interaction withdraws from that account. The math never changes: consistent small deposits over long periods outperform sporadic large ones.

Third: depth beats breadth. A handful of relationships characterized by genuine mutual investment will outperform a network of thousands characterized by efficient touchpoints.

The choice seems obvious when framed explicitly. Yet most professionals drift toward efficiency by default because optimization feels productive and relationship investment feels unmeasurable.

The strategic professional makes the opposite choice: automate the measurable work, and invest in the relationship work that creates disproportionate long-term value.

8

ELEVATE THE STRATEGIC

This is the final step. Not because the journey ends here but because this is where everything preceding it culminates into sustainable professional transformation. Steps one through four cleared the obstacles. You released the routine work draining your cognitive bandwidth. You expanded your creative capacity beyond what unaided cognition produces. You partnered with AI to accelerate analytical work while maintaining judgment authority. You protected the relationships that strategic work requires. Each step removed a constraint. This step applies the freedom those removed constraints create.

Strategic work becomes possible when you're not drowning in execution. Strategic thinking becomes sustainable when you have infrastructure supporting it. Strategic capability compounds when you systematically learn from every decision you examine rigorously. The professional renaissance you've been building toward requires this shift —from doing work to shaping work, from contributing output to directing outcomes, from executing strategy to creating it. The tools exist. The frameworks exist. The time exists because you've reclaimed it through the previous steps. What remains is the deliberate choice to build strategic capability through systematic practice that most professionals avoid because it requires admitting their current judgment might be less reliable than their confidence suggests.

Analysis is a process. Judgment is a capability.

Analysis involves gathering information, identifying patterns, evaluating options against criteria, and projecting outcomes—work AI accelerates dramatically. Judgment involves weighing incommensurable factors, making values-based trade-offs, reading human dynamics, and choosing when information is inherently incomplete—work that remains irreducibly human.

When you delegate analysis to AI, you're not delegating judgment. You're freeing judgment from the analytical burden that previously consumed its capacity. Better analysis enables better judgment by ensuring decisions rest on comprehensive data rather than whatever information happened to be at hand.

The strategic leader who tries to do all analysis personally has less cognitive capacity for judgment—the very capability that justifies their strategic role.

THE INTELLECTUAL SPARRING PARTNER

AI becomes most valuable at the strategic level not by providing answers but by improving your questions. The best strategic decisions emerge from rigorous examination of assumptions, consideration of alternative perspectives, and systematic evaluation of potential outcomes before committing resources.

Daniel Kahneman observed: "A reliable way to make people believe in falsehoods is frequent repetition, because familiarity is not easily distinguished from truth."

The observation applies directly to strategic thinking. You've made certain strategic assumptions so many times they feel true—but that sense of truth comes from repetition, not validity. The competitor you've underestimated for years. The market segment you've assumed isn't worth pursuing. The capability you've decided your organization can't build. These beliefs feel certain because they're familiar, not because you've recently tested them.

The adversarial frameworks—Red Team Protocol, blindspot detection, counter-argument generation—exist precisely to challenge beliefs that feel true because you've held them long. AI's value isn't that it's smarter than you. It's that it doesn't share your familiar assumptions. It questions what you've stopped questioning. It surfaces what your familiarity has hidden.

Most professionals make strategic decisions through predictable patterns: recognize the decision, gather available information, apply experience and intuition to evaluate options, choose based on what feels right given time constraints. This works adequately for routine choices. It fails systematically for non-routine strategic decisions where stakes justify deeper analysis.

AI enables systematic strategic rigor that mirrors how experienced strategists already think—but with computational thoroughness impossible for human cognition alone. You examine decisions from multiple angles simultaneously, test logical consistency of reasoning, and identify gaps in thinking before those gaps become costly mistakes.

THE RED TEAM PROTOCOL

The 'Red Team Protocol' transforms AI into your strategic adversary. Red teaming comes from military strategy—assigning a team to deliberately attack your plan to find weaknesses before enemies do.

Present your strategic plan to AI with explicit instructions: "Act as our primary competitor. You have full knowledge of this strategy. Your objective is to counter it effectively and capture our market position. Identify every vulnerability, assumption, and execution risk. Explain exactly how you would exploit each weakness."

The AI response reveals gaps invisible from inside your strategic perspective. Dependency on a single supplier you assumed was reliable. Market segment assumptions based on outdated customer research. Competitive response scenarios you hadn't considered. Technology shifts that could invalidate your approach. Each identified vulnerability becomes an opportunity to strengthen strategy before execution.

Structure red team sessions using the Vulnerability Assessment Framework:

Present your complete strategy including objectives, timeline, resource requirements, and success metrics.

Request AI to identify vulnerabilities in five categories—competitive response gaps, resource dependency risks, execution complexity issues, market assumption flaws, and timeline feasibility problems.

Document every identified vulnerability with specific descriptions.

Rate each vulnerability on likelihood and impact using a 1-5 scale.

Develop mitigation strategies for high-priority vulnerabilities before finalizing your strategic plan.

This systematic approach ensures comprehensive adversarial analysis rather than surface-level critique. A 45-minute red team session typically surfaces 8-15 significant vulnerabilities. Addressing even three or four major gaps before execution substantially improves strategic success probability.

For a market expansion strategy, AI red team analysis might surface: your customer acquisition cost assumptions rely on brand recognition that doesn't exist in new markets. Your pricing strategy assumes competitors won't respond aggressively when they almost certainly will. Your timeline assumes regulatory approval processes similar to your home market when they're actually three times longer. Your success metrics focus on revenue without accounting for regional profitability differences.

Apply red team analysis to every significant strategic decision. Budget allocations, organizational restructuring, product launches, partnership agreements, major hires. Spend 30 minutes having AI attack your strategy before you finalize and present it. The vulnerabilities AI identifies through adversarial analysis surface issues you would otherwise discover during execution—when they're expensive to fix.

BLINDSPOT DETECTION

Blindspot detection requires explicitly requesting what you're not seeing. Strategic blind spots emerge from three sources: information you don't have access to, perspectives you don't naturally consider, and assumptions so fundamental you don't recognize them as assumptions.

Prompt AI to identify potential blind spots by category. "What information might be missing from this analysis that could change the conclusion?" "What stakeholder perspectives haven't been adequately considered?" "What assumptions underpin this strategy that might not hold true?"

For a technology infrastructure decision, blindspot prompts might reveal: you're assuming current usage patterns will continue when rapid user growth could change access patterns entirely. You're evaluating vendors based on current capabilities without considering their development roadmap differences. You're optimizing for current team skills without accounting for the learning curve required for new systems. You're prioritizing immediate cost savings over long-term integration complexity.

Make blindspot analysis systematic. Before finalizing any strategic decision, run three explicit blindspot prompts focused on information gaps, stakeholder perspectives, and underlying assumptions. Document what each analysis reveals. Evaluate whether identified blind spots change your strategic choice or require additional investigation.

COUNTER-ARGUMENT GENERATION

Counter-argument generation strengthens your strategic position by forcing you to defend it rigorously. The quality of your strategic thinking shows most clearly not in your ability to argue for your preferred option but in your ability to articulate the strongest case against it.

Prompt AI to generate the most compelling argument against your strategic choice. Not strawman objections you can easily dismiss, but substantive challenges that force you to either strengthen your reasoning or reconsider your decision.

For an organizational restructuring proposal, counter-arguments might include: the productivity gains you project assume no transition disruption when restructuring historically creates 3-6 months of reduced productivity. Your efficiency analysis focuses on steady-state operations without accounting for knowledge loss from breaking up effective working relationships. Your cost savings projections don't include hidden coordination costs from new reporting structures. Your timeline assumes smooth implementation when organizational change typically encounters significant resistance.

Each counter-argument requires a response. Either you refine strategy to address the challenge, or you document why the counter-argument doesn't apply to your specific situation. Both outcomes strengthen decision quality—either through improved strategy or clearer reasoning about why your original approach remains optimal despite challenges.

Generate counter-arguments for every significant strategic recommendation before presenting to leadership. If you can't articulate and respond to the strongest case against your proposal, you haven't thought through the decision adequately.

COGNITIVE DIVERSITY SIMULATION

Cognitive diversity simulation helps you see strategic decisions through stakeholder perspectives you don't naturally adopt. The CFO sees financial implications you might miss. The head of operations sees execution challenges you might underestimate. The customers see value propositions differently than you do. The board sees competitive positioning through a different lens.

Prompt AI to analyze your strategic decision from specific stakeholder viewpoints. "Evaluate this strategy from the CFO's perspective, focusing on financial risk, resource allocation efficiency, and return on investment." "Analyze this from the customer's viewpoint, focusing on value received, alternatives available, and reasons to choose competitors instead."

For a pricing strategy decision, perspective analysis reveals different priorities. The CFO focuses on margin compression, revenue predictability, and competitive pressure on pricing power. The sales team focuses on deal velocity, win rates against competitors, and objection handling. The customer sees total cost of ownership, switching costs from current solutions, and risk of vendor dependence. The board focuses on market share growth versus profitability trade-offs.

Each perspective surfaces considerations you might deprioritize or miss entirely from your own vantage point. The CFO's analysis might reveal cash flow implications you hadn't considered. The customer perspective might identify friction points in your value proposition. The board viewpoint might highlight strategic positioning questions beyond immediate financial returns.

Run perspective analysis for every major stakeholder group affected by or involved in executing your strategic decision. Document what each perspective reveals. Adjust strategy to address legitimate concerns or strengthen your case for why certain trade-offs are necessary.

MULTI-VARIABLE ANALYSIS

Strategic decisions require balancing multiple variables—quantitative data, qualitative factors, short-term pressures, long-term goals, competing stakeholder interests, resource constraints, competitive dynamics, and organizational capabilities. Effective strategic thinking integrates these variables systematically rather than over-weighting whichever factor feels most salient in the moment.

AI enables multi-variable strategic analysis that maintains rigorous logic across complexity that quickly overwhelms human working memory. Working memory holds only four to seven pieces of information simultaneously. Strategic decisions often involve dozens of relevant factors. AI helps you maintain analytical rigor across all variables rather than simplifying prematurely to make the decision cognitively manageable.

Multi-variable analysis requires explicitly defining all decision factors and their relationships. Most strategic frameworks prioritize certain variables—financial returns, customer satisfaction, operational efficiency, competitive advantage—while treating other considerations as secondary. This prioritization makes sense for routine decisions within stable contexts. It fails for non-routine strategic choices where the "secondary" factors might be decisive.

Create comprehensive decision frameworks that map all relevant variables and their interdependencies. Use AI to help structure the analysis by providing your decision context and requesting a multi-variable framework.

The Decision Matrix Framework organizes complex choices systematically:

List all decision factors relevant to your specific choice—both quantitative metrics and qualitative considerations.

For each factor, assign importance weight from 1-10 based on your strategic priorities.

Evaluate each option against every factor using a consistent scale.

Calculate weighted scores showing how options perform across all considerations.

Analyze trade-offs explicitly—where choosing option A means accepting lower scores on factors B and C.

For a vendor selection decision, comprehensive analysis includes: total cost of ownership over three years including hidden costs, implementation timeline and resource requirements, feature fit against current and projected needs, integration complexity with existing systems, vendor financial stability and roadmap, support quality and response times, security and compliance implications, switching costs if the relationship doesn't work out, organizational change management requirements, and strategic flexibility for future growth.

Prompt AI to evaluate each variable systematically. Provide relevant data where available. Request analysis of trade-offs between variables —where optimizing for cost might sacrifice flexibility, or prioritizing features might increase implementation complexity, or emphasizing vendor stability might mean accepting less innovative solutions.

The analysis won't make the decision for you. It will clarify the decision structure so you understand exactly which trade-offs you're making and why. When you choose the vendor with higher costs but lower implementation risk, you're making that choice deliberately based on current organizational capacity for complex projects—not defaulting to lowest price because cost was the most visible variable.

Apply this framework to every significant strategic decision—those affecting substantial resources, involving multiple stakeholders, or creating commitments lasting six months or longer.

Document multi-variable analysis for every strategic decision above a certain threshold. The specific threshold depends on your role and organization—perhaps decisions affecting $50,000 or more, or decisions with six-month or longer implications, or decisions requiring cross-functional coordination.

THE PRE-MORTEM TECHNIQUE

The Pre-Mortem technique helps you identify failure modes before committing to execution. Traditional project post-mortems analyze what went wrong after failure. Pre-mortems assume failure then work backward to identify what caused it—surfacing risks that forward-looking planning often misses.

Prompt AI to conduct a pre-mortem: "Assume this strategy failed completely six months after implementation. Describe the most likely sequence of events that led to failure. Identify early warning signs we should have noticed. Explain what we should have done differently."

For a product launch strategy, pre-mortem analysis might reveal: we assumed market readiness based on survey responses, but actual purchase behavior at launch showed the market needed more education than we provided. We optimized the product for features customers said they wanted, but they actually made purchase decisions based on ease of implementation which we underemphasized. We planned launch timing around our development schedule rather than market buying cycles, launching right before Q4 budget freezes made purchasing difficult. We built minimal customer support infrastructure assuming adoption would be gradual, but unexpected viral growth overwhelmed our capacity and damaged early customer experience.

Each pre-mortem scenario becomes a risk to mitigate before launch. Either adjust strategy to prevent the failure mode, or create contingency plans for rapid response if early indicators emerge. The pre-mortem doesn't predict the future. It systematically surfaces failure scenarios that forward-looking planning overlooks because success feels inevitable when you're designing the strategy.

Run pre-mortems for every significant strategic initiative before finalizing the execution plan. Time-box the analysis to 45 minutes. Document the identified failure modes and corresponding mitigation strategies. Review the pre-mortem analysis monthly during execution to check whether early warning indicators have appeared.

OPTION GENERATION

Option generation prevents strategic tunnel vision by forcing consideration of alternatives beyond the obvious choices. Most strategic decisions present as binary choices—expand into market A or market B, hire for role X or role Y, implement technology 1 or technology 2. This binary framing excludes creative alternatives that might be superior to both default options.

Prompt AI to generate five distinct strategic approaches to your objective, explicitly requesting diversity. "Generate five different strategies for achieving [objective], where each approach is fundamentally different in its core mechanism, not just variations of the same basic strategy."

For an objective of increasing customer retention by 20%, diverse strategies might include: intensive relationship management for top 20% of accounts, complete product experience redesign focusing on first 90 days, loyalty program with tiered benefits, strategic pricing restructure making switching costs more visible, customer success technology platform enabling proactive intervention, community building initiatives creating network effects, or partnership ecosystem integration making your product sticky through third-party dependencies.

Each option represents a different strategic theory about what drives retention. The relationship management approach assumes personal connection matters most. The experience redesign assumes product friction causes churn. The loyalty program assumes economic incentives drive behavior. The technology platform assumes data-driven intervention works. The community approach assumes peer relationships create commitment. The ecosystem strategy assumes integration lock-in prevents switching.

Evaluate all options against your specific context. Most won't be optimal for your situation. The process ensures you're choosing the best strategy through systematic evaluation rather than implementing the first plausible idea. Often the final strategy combines elements from multiple generated options—relationship management for tier one accounts, experience improvements for everyone, and technology infrastructure enabling both.

Generate options for every strategic challenge before committing resources. Request five distinct approaches. Evaluate each against decision criteria. Document why you selected your chosen strategy and why you rejected alternatives. This documentation matters when circumstances change and you need to revisit the decision—you'll understand why alternatives were initially rejected and whether those reasons still hold.

THE DECISION JOURNAL

Strategic capability improves through systematic learning from past decisions—analyzing what happened, understanding why, and applying those insights to future choices. Most professionals learn sporadically from obvious successes and failures while missing subtler patterns that would improve strategic judgment more significantly.

Research on judicial decisions demonstrates how dramatically cognitive depletion affects judgment quality. A study in the Proceedings of the National Academy of Sciences found that judges granted parole in about 65% of cases at start of day and immediately after breaks, but approval rates dropped to nearly 0% just before breaks—when decision fatigue peaked. Same case, same judge, same evidence—radically different outcomes based on when the decision occurred.

Strategic decisions are similarly vulnerable. The allocation choice you make at 4:30 PM after eight hours of meetings isn't the same decision you'd make at 9:00 AM with cognitive reserves intact.

A decision journal transforms experience into systematic strategic learning by documenting decision reasoning at the moment of choice, tracking outcomes over time, analyzing results against expectations, and identifying patterns that refine future judgment. The journal creates a feedback loop between decisions and outcomes that accelerates strategic skill development.

DECISION DOCUMENTATION

Decision documentation captures your reasoning process and predictions at time of choice. Memory reconstructs past reasoning to match present reality—you remember anticipating outcomes that actually surprised you, or you forget concerns you initially had about decisions that worked out well. This retrospective distortion prevents learning from actual decision quality versus outcome luck.

Document strategic decisions using a standard template immediately after making the choice, before execution begins. The Strategic Decision Record template includes seven essential components capturing your complete reasoning process.

Record the decision context including the problem or opportunity, stakeholders involved, time constraints, and available resources.

Document all alternatives you considered with brief descriptions of each option and why you evaluated them.

Specify your chosen option with detailed rationale explaining why this choice beats alternatives based on your analysis.

List key assumptions underlying your decision—market conditions you're counting on, organizational capabilities you're assuming, competitive responses you're predicting, and resource availability you're banking on.

Define expected outcomes with specific metrics and timeframes—not vague goals like "improve performance" but concrete targets like "reduce customer acquisition cost by 15% within six months" or "achieve 25% market share in the SMB segment by Q4."

Identify anticipated risks with specific mitigation strategies for each—don't just list what might go wrong, document what you'll do if it does.

Include your confidence level on a 1-10 scale for calibration tracking.

This systematic documentation requires 30-45 minutes per significant decision but transforms scattered experience into structured learning.

When documenting a strategic decision, capture eight key elements:

First, the decision itself—what exactly you're deciding to do.

Second, the context—why this decision matters right now and what prompted it.

Third, the options you considered—what alternatives you seriously evaluated before making this choice.

Fourth, your chosen approach—what you're doing and the reasoning behind it.

Fifth, your key assumptions—what you're counting on being true for this decision to work.

Sixth, the risks you're accepting—what could go wrong that you're choosing to proceed despite.

Seventh, your success criteria—how you'll know in six months whether this decision actually worked.

And eighth, your confidence level—rate yourself on a scale of one to ten for how certain you feel about this choice.

Store this document somewhere you'll rediscover it in six months.

Here's what this looks like in practice. For a hiring decision, your journal would include: the role requirements and specific team gaps this hire addresses; your evaluation of the candidate across both technical skills and cultural fit; the alternative candidates you considered and why you didn't select them; the compensation and equity structure with your justification; the expected performance impact over the first year, including specific deliverables you anticipate; your onboarding plan with clear success metrics; and finally, the retention risks you've identified along with your mitigation approach.

The documentation requires 30-45 minutes per significant decision. This investment seems expensive until you calculate the cost of poor strategic choices. A single bad senior hire costs six months of productivity and $150,000 in compensation and opportunity cost. Spending 45 minutes documenting decision reasoning is cheap insurance against repeating the same hiring mistakes.

Store decision documentation in searchable format accessible to future you. As your strategic responsibilities grow, you'll make thousands of decisions. The journal helps you track which types of decisions you make well consistently and which types show repeated errors despite experience.

OUTCOME TRACKING

Outcome tracking connects decisions to results over appropriate timeframes. Strategic decisions often show results months or years after execution begins. Without systematic tracking, you lose the connection between choice and outcome—undermining your ability to learn what works in your specific context.

Create a quarterly review process examining decisions made 3-12 months earlier. Compare actual outcomes against documented predictions. Categorize results as better than expected, matching expectations, worse than expected, or too early to evaluate. For each category, analyze why.

Outcomes better than expected reveal opportunities. Did you underestimate your capabilities? Did market conditions improve unexpectedly? Did the strategy work better than theory suggested? Understanding positive surprises helps you recognize when to be more ambitious or which approaches deserve more investment.

Outcomes matching expectations validate your strategic judgment for that decision type. Document the reasoning process that worked so you can replicate it for similar future choices. These are your strategic competencies—decision categories where your judgment consistently produces predicted results.

Outcomes worse than expected identify learning opportunities. Did you miss information that was available? Did you overestimate organizational capabilities? Did you fail to account for competitor responses? Did execution problems undermine sound strategy? Distinguishing between strategic errors and execution failures matters—they require different solutions.

Decisions still too early to evaluate require continued tracking. Document which outcomes you're waiting to observe and when you'll next review. Some strategic choices take years to show full results. The journal prevents premature conclusions while ensuring you eventually learn from every significant decision.

PATTERN RECOGNITION

Pattern recognition across multiple decisions builds strategic wisdom faster than experience alone. Individual decisions teach specific lessons. Patterns across decisions reveal systematic strengths and blind spots in your strategic thinking.

Analyze decision journal entries quarterly to identify patterns. Group decisions by type—hiring, budget allocation, vendor selection, strategic partnerships, organizational changes. For each category, examine: how accurate your outcome predictions are, which risks you consistently anticipate and which you miss, what assumptions frequently fail, and where execution diverges from plans.

The Pattern Analysis Protocol creates systematic learning from accumulated decisions:

Every quarter, review all decisions made 3-12 months prior where outcomes are now visible.

Group decisions into categories based on decision type and complexity level.

Calculate accuracy rates for your predictions in each category—how often did outcomes match your expectations?

Identify recurring failure modes—what types of risks do you consistently miss?

Spot assumption patterns—which assumptions prove reliable and which frequently fail?

Compare confidence levels to actual results—does high confidence correlate with better outcomes?

Pattern analysis might reveal: you consistently underestimate implementation timelines for technology projects by 40%, your hiring decisions for creative roles work out well but analytical role hires frequently disappoint, budget allocation decisions focused on cost optimization succeed but growth investment decisions show mixed results, and vendor selections based on relationship quality outperform feature-based selections.

These patterns indicate where you have strong strategic judgment and where systematic biases exist. For areas of strength, document what you're doing right so you can teach others and maintain quality as responsibilities grow. For areas showing systematic errors, design corrective processes—perhaps external validation for hiring decisions in your weak areas, or additional contingency time for technology implementations, or different evaluation criteria for vendor selection.

Use AI to help analyze patterns by feeding it decision documentation and outcomes. Prompt for pattern identification across decision categories. AI analysis might surface trends you miss: decisions made under time pressure show 30% worse outcomes than decisions with adequate evaluation time, decisions where you consulted with domain experts succeeded 65% more often than decisions made independently, or decisions aligned with your core expertise show much better outcomes than decisions in adjacent domains.

INTUITION CALIBRATION

Intuition calibration helps you understand when to trust your gut and when to demand more analysis. Strategic intuition develops through pattern recognition across many experiences. Some professionals develop well-calibrated intuition that produces reliable judgment. Others maintain persistent blind spots despite extensive experience. The decision journal reveals which type you are and for which decision categories.

Calculate calibration by comparing intuitive confidence to actual outcomes. For each documented decision, note your confidence level on a 1-10 scale. Track whether decisions where you felt confident actually succeeded more often than decisions where you were uncertain. Well-calibrated intuition shows correlation—high confidence decisions succeed more frequently than low confidence decisions. Poor calibration shows weak or negative correlation—you feel equally confident about decisions that succeed and decisions that fail.

If your intuition shows good calibration for certain decision types, trust it more for those categories. If calibration is poor, require more systematic analysis and external validation before committing to choices based on how they feel.

Most professionals show better intuition calibration in their areas of deep expertise and poorer calibration in adjacent domains where they have enough experience to feel confident but not enough to develop accurate pattern recognition. The decision journal reveals these boundaries explicitly, helping you recognize when you're operating inside versus outside your zone of reliable judgment.

AN EXAMPLE

Beth directed admissions for a mid-sized state university—managing a team of five counselors, overseeing 12,000 applications annually, dealing with an inbox that never stopped. She knew she was supposed to be thinking strategically about enrollment positioning, competitive differentiation, and yield optimization. Instead, she spent days reviewing application materials, coordinating campus visit logistics, and managing the internal politics of academic department relationships.

Her first strategic AI experiment was the Red Team Protocol

applied to her fall recruitment plan. She'd already presented the plan to her provost—but something bothered her about the October campus event series. She prompted AI to attack the plan from her primary competitor university's perspective.

The vulnerability surfaced immediately: her October open house schedule conflicted directly with early decision deadlines at three competing universities, when prospective students would be finalizing applications there. Her events would compete for attention during the exact window when admitted students at rival institutions were making commitment decisions. The AI identified what her planning process had missed because she'd focused on her campus calendar, not the competitive enrollment landscape.

She restructured the fall calendar, moving signature events earlier to capture attention before competitor deadlines locked in student attention. When those November deadlines arrived, her university was already top-of-mind—positioned as the choice students were comparing others against, not the backup option they were considering after rejection.

That single Red Team session changed her relationship with strategic thinking. She began running Pre-Mortems before major recruitment initiatives, identifying enrollment risks she would have discovered only when applications fell short. She started the Decision Journal for every significant budget allocation, building a track record that revealed patterns in her judgment—where her instincts about student behavior were consistently right, and where she needed data validation.

Eighteen months later, Beth's director role had transformed. The tactical work—application reviews, event coordination, meeting management—now took two hours per day instead of six. The remaining four hours went toward the strategic work her role was supposed to contain: competitive enrollment analysis, relationship development with high school counselors, and the market sensing that identified demographic shifts before they became obvious.

Her team noticed the shift. Her provost commented that she'd "become more strategic." Her president started including her in budget conversations she previously would have been informed about afterward.

What changed wasn't Beth's capabilities—she'd always had a strategic instinct about enrollment. What changed was how she allocated attention. AI-augmented analysis freed cognitive capacity from data processing grunt work. The frameworks gave structure to strategic thinking she'd done episodically. The combination moved her from tactical execution to strategic leadership—exactly what her director role required, and what she'd never had bandwidth to provide.

STRATEGIC AUTHORITY

The shift from execution to strategy changes how others perceive your professional value. You move from being evaluated on output volume—how many reports you write, how many analyses you complete, how many projects you deliver—to being evaluated on outcome quality—whether your strategic recommendations were correct, whether your decisions created value, whether your judgment proved sound.

Strategic leaders who master AI-augmented decision-making report consistent transformation: they shift from reactive to proactive positioning. The time previously spent gathering information, conducting analysis, and researching alternatives now goes toward anticipating challenges before they materialize, building relationships that create future options, and positioning resources for opportunities not yet visible.

This shift changes how others perceive your strategic contribution. Reactive leaders are problem-solvers—valuable, but responding to circumstances others have created. Proactive leaders are opportunity-creators—shaping circumstances that others then respond to. The distinction affects everything from how you're viewed in leadership discussions to whether you're consulted before decisions are made or informed after.

The same cognitive capacity that AI frees from analytical work becomes available for the forward-looking, relationship-building, opportunity-sensing work that distinguishes strategic leaders who shape outcomes from operational managers who execute within them.

THE ARCHITECT MINDSET

The architect mindset frames your evolving role clearly. Strategic professionals design and direct while others execute detailed work. The architect doesn't personally install electrical systems or pour concrete, but they're responsible for whether the building stands and serves its purpose. They set the vision, make critical design choices, ensure components integrate correctly, and verify quality against specifications.

Apply the architect mindset to your strategic work. You're responsible for outcomes but you don't personally complete every tactical task required for success. You define what success looks like, establish evaluation criteria, allocate resources appropriately, coordinate efforts across teams, verify execution quality, and adjust course when needed.

This doesn't mean disconnection from implementation. Architects visit construction sites regularly to verify work matches designs. Strategic professionals maintain sufficient connection to execution to catch problems early, understand constraints that affect future planning, and maintain credibility with execution teams. But you're not personally doing the execution work except in specific situations where your expertise is irreplaceable.

The time AI saves you through automation of routine work and acceleration of analytical work doesn't go to doing more execution faster. It goes to strategic work—scanning for emerging risks and opportunities, building relationships that create future options, developing frameworks that improve decision quality, and setting direction rather than following it.

LEADERSHIP THROUGH AI

Leadership through AI requires you to set direction while AI maps the route. AI helps evaluate options, analyze trade-offs, model scenarios, and surface considerations you might miss. But AI cannot determine what matters to you, what values you're unwilling to compromise, which stakeholders deserve priority, or what future you're trying to create.

The most valuable strategic questions AI cannot answer for you: "Should we prioritize revenue growth or profitability this year?" "Should we enter market A where we have relationships or market B where we see opportunity?" "Should we maintain current strategic focus or pivot toward emerging trends?" "Should we optimize our current business model or invest in disruptive innovation?"

These questions require judgment based on values, vision, context, and interpretation of ambiguous information. AI helps you explore the implications of each choice thoroughly. It models scenarios, surfaces trade-offs, and ensures you understand consequences. But the choice remains yours.

This is the essence of strategic leadership—making decisions with incomplete information under uncertainty where being wrong has significant consequences. The responsibility cannot be delegated to AI or anyone else. It's why strategic roles command premium compensation. It's why organizations pay for senior judgment.

The subtle trap: letting AI influence not just what you analyze, but what you decide. The Red Team Protocol should surface vulnerabilities for your judgment to evaluate—not determine which vulnerabilities are important. The Pre-Mortem should generate failure scenarios for your assessment—not prioritize which failures to prevent. The Decision Journal should document reasoning for your reflection—not prescribe what you should have reasoned.

You'll recognize you've crossed this line when AI recommendations start feeling authoritative rather than informative. When you're reluctant to decide without AI input. When you find yourself justifying decisions by reference to AI analysis rather than by your own judgment of what that analysis means.

The correction is simple: after any AI-assisted strategic analysis, close the tool and sit with your own thinking for at least ten minutes before deciding. Let your judgment engage with the analysis rather than be replaced by it. If your conclusion differs from what AI analysis seems to suggest, understand why—and trust your contextual understanding when it's grounded in information the AI doesn't have.

AI extends your analytical capacity. It doesn't substitute for your judgment capacity. Confusing the two undermines the very capability that makes you strategically valuable.

THE VISIONARY ROLE

The visionary role centers on capabilities AI lacks fundamentally. AI cannot dream about possibilities that don't exist in its training data. It cannot set goals based on personal values. It cannot inspire people to pursue difficult objectives. It cannot build the trust required for teams to execute effectively under pressure. It cannot provide the moral and emotional leadership that organizational challenges demand.

Your strategic value concentrates in these irreplaceable human capabilities as AI handles more analytical and execution work. Vision requires imagining futures that don't exist yet and articulating why those futures matter enough to justify present sacrifice. This is human work—connecting abstract goals to emotional meaning, painting vivid pictures of possibility, and giving people reasons to believe difficult achievements are worthwhile.

Strategy requires choosing among incompatible goods—growth versus profitability, innovation versus execution, flexibility versus efficiency, speed versus quality. These trade-offs have no objectively correct answers. They require judgment about what matters most in your specific context given your values and vision. AI helps you understand the implications of each trade-off more thoroughly. The judgment remains yours.

Leadership requires building psychological safety, earning trust, demonstrating competence under pressure, and making people believe they can achieve more than they thought possible. These are fundamentally human social activities. AI cannot replicate the emotional connection and authentic presence that effective leadership requires.

Focus your strategic development on these capabilities AI cannot replicate. Practice articulating vision in ways that resonate emotionally. Develop your judgment through systematic decision-making with feedback. Build relationships based on authentic connection and demonstrated reliability. These skills become more valuable as AI handles more technical and analytical work—creating divergence between professionals who invest in human capabilities and those who focus on technical skills AI will eventually match or exceed.

THE COMPLETE ARC

The four previous steps—Release, Expand, Partner, Protect—cleared the space for strategic work by automating routine tasks, amplifying creative capacity, accelerating analytical work, and preserving relationship quality. Elevate puts that space to use by focusing your energy on work that requires strategic judgment, visionary thinking, and authentic leadership.

Apply the five steps systematically over 60 days. By day 60, you've built the infrastructure for sustained professional renaissance. Your relationship with work fundamentally changes. The anxiety that prompted this journey—the fear of obsolescence, the sense of falling behind, the worry about professional relevance—transforms into confidence based on demonstrated capability for strategic adaptation.

You're not competing with AI. You're directing it toward objectives you define based on judgment it cannot replicate. The routine work no longer consumes your energy. The creative work expands your capacity. The analytical work accelerates your insight. The relationships deepen your influence. The strategic work elevates your impact.

This is the professional renaissance. Not survival through technological adoption. Flourishing through strategic amplification of irreplaceable human capabilities.

9

MASTERING THE DANCE

Competence becomes its own trap. You master something—a skill, a system, a way of working—and that mastery creates comfort. The comfort creates attachment. The attachment creates resistance to change. Then change happens anyway, and suddenly your hard-won competence feels like an anchor rather than an advantage. This is the central paradox of AI integration: the better you get at working with today's AI capabilities, the more difficult it becomes to adapt when tomorrow's capabilities arrive. The delegation system you've built works beautifully. Your custom prompts produce consistent results. Your workflow automation saves hours weekly. Then a model update changes how the AI responds. A new competitor emerges with different capabilities. Your industry discovers applications you hadn't considered. The dance changes, and your carefully choreographed routine no longer matches the music. This chapter isn't about preventing that disruption—that's impossible. It's about developing the specific capabilities that let you adapt gracefully rather than starting over from scratch every time the technology shifts.

DYNAMIC LEARNING STRATEGIES

The average knowledge worker receives 47 AI tool announcements monthly through newsletters, LinkedIn, colleague recommendations, and platform notifications. Following every update requires 8-12 hours weekly just to stay informed, before any implementation or testing.

You need a filtering system.

The Just-in-Time Learning Model operates on one principle: learn the tool when you have the problem it solves, not before. This inverts traditional professional development where you acquire skills proactively then look for applications. That model works for stable technologies with long learning curves. It fails for rapidly evolving tools with short implementation cycles.

Apply just-in-time learning by maintaining problem-first orientation. When you encounter a workflow bottleneck, research limitation, or creative constraint, investigate whether AI tools address it. Not when someone mentions a new platform. Not when you see a tutorial. When you have an actual problem requiring an actual solution.

Cognitive science supports this approach. Studies on the spacing effect demonstrate that skills applied shortly after acquisition show dramatically higher retention than skills learned without immediate use. Research in *Psychological Science in the Public Interest* found application-based learning produces retention rates four to five times higher than passive study. Learning in response to actual problems isn't just convenient—it's neurologically optimal.

Three practical advantages: First, you learn with immediate application context, accelerating retention. Second, you avoid accumulating theoretical knowledge about tools you'll never use—most professionals do this extensively, creating cognitive overhead without value. Third, you develop genuine assessment skills rather than relying on marketing claims, because you're evaluating tools against real performance requirements.

Implement just-in-time learning through the Three-Question Filter. Before investing time learning any new AI tool or feature, answer three questions: What specific problem am I solving right now? How would this tool solve that problem in ways my current approach cannot? What's my backup plan if this tool doesn't work as expected?

If you can't answer all three concretely, defer learning until you have a clearer application.

Filter bubble curation determines what information reaches your attention. Without deliberate curation, you default to whatever shows up in feeds—hype, speculation about future capabilities, and technical detail inappropriate for strategic users.

You need three types of sources, each serving distinct purposes.

Practitioner sources focus on real implementation in contexts similar to yours. Professionals using AI tools to solve actual problems in day-to-day work. They discuss what works, what doesn't, actual learning curves. They share prompts that produce results, not theoretical possibilities.

Identify 5-8 practitioners whose work context resembles yours—same industry, similar role, comparable organizational complexity. Follow their updates specifically. When they recommend something, investigate. When they criticize something, take it seriously.

Signal sources filter noise and highlight genuinely significant developments. Technical professionals who understand AI capabilities deeply but communicate for strategic audiences. They distinguish incremental improvements from fundamental capability shifts. They explain why certain developments matter for professional practice while dismissing hype.

Find 2-3 signal sources you trust. These might be technology analysts, AI consultants, or researchers who translate work for practical application. Check weekly, not daily.

Implementation sources provide tactical guidance for specific tools and workflows. Tutorial creators, documentation writers, workflow specialists who explain exactly how to accomplish specific tasks.

You don't follow these continuously. You search for them when you need to solve particular problems. Bookmark 3-5 high-quality implementation sources you can reference when needed.

The 5-8-3 Curation Formula: Follow five to eight practitioners, two to three signal sources, and bookmark three to five implementation resources. When a practitioner recommends a tool, investigate. When they share a workflow, consider adapting it. When they criticize something, take the criticism seriously. Check signal sources weekly to calibrate your understanding of what's actually changing. Search implementation sources only when you have specific problems to solve.

Curate for quality and relevance, not comprehensiveness. You're not trying to know everything about AI. You're staying informed about developments relevant to your strategic position.

The update cycle challenge requires distinguishing hype from meaningful capability shifts. Major model updates generate enormous attention, but not all justify changing workflows. Some improve speed or accuracy marginally. Others unlock entirely new capabilities reshaping what's possible.

Evaluate updates using the Capability Threshold Framework. New capabilities enabling tasks previously impossible or impractical justify workflow revision. Improvements to existing capabilities justify testing but rarely require immediate revision unless improvement is dramatic —typically 50% or greater performance gain.

When AI image generation crossed from experimental to production-ready quality, that justified workflow revision for visual professionals. When text generation improved from GPT-3.5 to GPT-4, enhanced reasoning capability justified testing for analytical work. When speech-to-text accuracy improved from 85% to 95%, that crossed the threshold for reliable transcription use.

But when a model becomes 10% faster at completing the same tasks? Nice but doesn't change your strategic approach. When interface improvements make a tool slightly more convenient? Appreciate it but don't rebuild workflows. When a new feature promises benefits but requires significant learning investment? Defer until you have a clear application need.

Most model updates don't justify workflow revision. Processing speed improves marginally. Accuracy increases slightly on specific task types. Interface elements shift. Pricing structures adjust. New features launch in beta. These are incremental improvements—pleasant upgrades requiring no adaptation.

The updates justify attention cross capability thresholds: enabling entirely new task types, dramatically improving previously unreliable capabilities, or introducing fundamentally different interaction paradigms.

Treating incremental updates with threshold-update urgency creates constant disruption without proportional benefit. Your quarterly review suffices for evaluating whether incremental improvements have accumulated to threshold-crossing significance. Between reviews, continue with your working approach.

Create a quarterly update assessment process. Every three months, spend 90 minutes reviewing major AI developments from the previous quarter. Identify which developments represent threshold capability shifts versus incremental improvements. For threshold shifts, plan specific testing within your workflow. For incremental improvements, note them but defer action unless clear application emerges.

Continuous calibration ensures automation-amplification decisions remain current as AI capabilities expand. Work you classified "human only" last year might be "AI-assisted" this year. Tasks designated for full automation six months ago might now benefit from human-AI partnership as models improve at handling nuance.

Schedule systematic re-audits of work categorization every three months. This doesn't mean questioning every delegation decision—most remain valid. Technology changes create opportunities to refine boundaries between automated, amplified, and human-only work.

Use the Re-Calibration Audit Framework quarterly. Return to your original cognitive energy audit—the categorization of tasks across Robot Zone (automate), Genius Zone (protect), and Gray Zone (partner). For each category, ask: Have AI capabilities improved enough to shift any tasks to different categories? Have I discovered limitations in AI handling of tasks I initially automated? Have I identified opportunities to amplify work I'm still handling manually?

Common re-calibration patterns: Tasks requiring careful human review six months ago might now be reliable enough for spot-checking rather than comprehensive review. Creative work needing human first drafts might now work better with AI first drafts that humans refine.

Analysis requiring complete human execution might now benefit from AI preprocessing that humans verify and interpret.

Calibration isn't just about what AI can do—it's about what you've learned about effective partnership. You might discover certain tasks work better with AI assistance than initially expected, not because technology improved but because you developed better collaboration skills. Or you might find tasks automated too aggressively, where restoring human judgment improves outcomes.

Document re-calibration decisions. Create simple log tracking what changed, why you made the change, and what results you observed. This documentation helps identify patterns in your own learning and capability development over time.

Tool fatigue relapse happens even to professionals with strong AI practices. The overwhelming feeling that there's too much to track, too much changing, too much demand on attention.

Relapses typically trigger during high-stress periods when bandwidth is already consumed, immediately after major AI announcements generating floods of commentary, when you encounter tasks your current system handles poorly and start questioning everything, or during comparison moments when colleagues seem more capable or current.

The pattern: anxiety about missing something important leads to expanded information consumption, which leads to overwhelm, which leads to either frantic tool-chasing or complete disengagement. Neither response serves you.

The intervention: return to your curation filters. Check your 3-5 trusted practitioner sources. If none are emphasizing the thing creating anxiety, it probably doesn't merit attention. If they are emphasizing it, investigate specifically that one thing. The filter system handles anxiety better than trying to process raw information flow yourself.

Tool fatigue relapse is normal. Having a return-to-baseline protocol transforms temporary destabilization into manageable blips rather than system-destroying spirals.

FEEDBACK LOOPS

Strategic AI integration requires systematic evaluation of whether your approach actually works. Most professionals operate on intuition—this feels more efficient, that seems to save time, I think quality holds steady. Intuition helps, but measurement prevents self-deception.

Personal performance metrics track whether AI integration delivers intended outcomes. Three categories: time recovery, output quality, and professional satisfaction.

Time recovery metrics measure whether you're actually reclaiming hours, not just working faster while maintaining the same total workload.

Track time spent on routine tasks weekly. Compare against the baseline from your initial cognitive energy audit. Track time available for strategic work—hours for deep thinking, relationship building, creative problem-solving. Compare against baseline. Track total working hours to ensure efficiency gains don't simply enable longer workdays.

For most professionals implementing Release-Expand-Partner-Protect-Elevate systematically, time recovery should show 8-15 hours weekly reclaimed from routine work within three months. If you're not seeing similar recovery, either delegation isn't comprehensive enough or reclaimed time is being consumed by new routine work rather than redirected to strategic activities.

Output quality metrics verify that delegation and amplification maintain or improve work quality rather than degrading it. Specific metrics depend on your role—response time for client communications, accuracy of analysis, effectiveness of creative work, strategic decision outcomes.

Choose 3-5 concrete quality indicators relevant to your work. For consultants: client satisfaction scores, proposal win rates, project outcome success. For marketers: campaign performance metrics, content engagement rates, creative asset effectiveness. For analysts: recommendation adoption rates, prediction accuracy, stakeholder feedback on insights provided.

Track monthly. Compare against pre-AI baseline. Quality should either improve or remain stable. If quality declines, investigate whether you've automated too aggressively, whether you need to refine how you partner with AI, or whether you need to restore more human involvement in specific workflow components.

Professional satisfaction metrics assess whether AI integration improves your experience of work or just makes you more productive at cost of engagement.

Track three subjective measures weekly using a simple 1-10 scale: energy level—how energized versus depleted you feel about work, engagement—how interested and absorbed you are in work, and confidence—how assured you feel about professional capabilities and future prospects.

These subjective measures matter as much as objective productivity gains. If you're reclaiming 10 hours weekly but feeling less engaged with work, something's wrong. The goal isn't just efficiency—it's elevating professional experience while maintaining effectiveness.

Record metrics in whatever tracking system you already use. A simple spreadsheet with weekly entries takes three minutes and provides data patterns over time. The key is consistency, not precision.

Quality audit protocols verify that AI-assisted output maintains your standards. This becomes especially important as delegation becomes routine—risk of drift increases as you stop reviewing carefully because "AI usually gets it right."

Implement systematic sampling audits for any work category where AI handles significant components. For routine communications, audit 10% of AI-drafted messages monthly—reading them with fresh eyes as if you're the recipient, not the sender. For analytical work, audit 20% of AI-processed analysis monthly—verify that conclusions are supported, logic is sound, nuance is preserved. For creative output, audit 15% of AI-amplified content monthly—ensure voice remains consistent, quality meets standards, human judgment is appropriately integrated.

The audit isn't about finding problems—though you will. It's about maintaining awareness of what your integrated workflow actually produces, not what you assume it produces. Small quality issues compound over time if undetected. Regular auditing catches drift before it becomes embedded in practice.

When audits identify quality issues, diagnose the root cause. Is AI capability insufficient for this task type? Is your prompting strategy unclear? Are you reviewing too quickly without adequate attention? Is the task actually in your Genius Zone and shouldn't be delegated at all? Each root cause suggests different corrections.

External feedback mechanisms prevent blind spots about how others experience your AI-augmented work. Your assessment of quality might differ from stakeholders' perception. What you consider acceptable AI-assisted communication might feel impersonal to clients. Analysis you view as sufficient might lack depth colleagues expect.

Create explicit feedback channels for key stakeholders. For close collaborators, schedule quarterly check-ins specifically asking: "Have you noticed any changes in my work quality, responsiveness, or communication style over the past three months? Are there areas where you'd like to see adjustment?"

Frame this openly. Don't hide AI use—most professionals in knowledge work now use AI to some degree, and transparency builds trust more than secrecy. You're not asking permission. You're gathering data about external perception of work quality.

For clients or other important relationships, incorporate feedback questions into existing touchpoint conversations. Don't create special "AI feedback" sessions that feel awkward. Add questions to regular project reviews or relationship check-ins: "How has our communication been working for you?" "Is the analysis I provide hitting the right level of detail?" "Are there aspects of how we work together you'd like me to adjust?"

Pay attention to indirect signals. Response time from stakeholders when you send messages—are they engaging as readily as before? Meeting attendance when you schedule—are people showing up or declining more often? Engagement with your output—are people reading, commenting, implementing, or has interaction quality decreased?

These indirect signals often reveal quality perception issues before anyone explicitly raises concerns. If you notice declining engagement, investigate whether AI assistance has inadvertently affected work quality, communication tone, or relationship maintenance.

Optimization cycles refine AI integration based on accumulated experience. After three months of systematic implementation, you have data about what works well, what needs adjustment, and what doesn't work at all in your specific context.

Schedule formal optimization reviews every three months. Review performance metrics, audit findings, and external feedback. Identify three categories: what's working well and should be expanded, what needs refinement and how to improve it, and what isn't working and should be discontinued or significantly revised.

Build a prompt library for tasks you've optimized successfully. When you discover a prompt structure consistently producing good results, save it. Document the context where it works, refinements that improved it, and any limitations you've identified. This library becomes increasingly valuable over time—you're building a custom toolkit matched to your specific work patterns.

Optimize workflows based on bottleneck identification. Where does AI integration create friction or delay? Where do you find yourself working around the AI-assisted approach rather than with it? These bottlenecks indicate opportunities for workflow revision or task reclassification.

The optimization mindset recognizes that your first implementation attempt won't be optimal. You're learning both what technology can do and how you work best with it. Systematic optimization cycles help you refine both dimensions progressively.

ADVANCED INTEGRATION

Basic AI integration focuses on task-level implementation—using AI to handle specific activities within existing workflows. Advanced integration involves workflow-level and system-level changes that reshape how you approach entire categories of work.

Custom GPTs enable specialized AI assistants trained on your specific context, terminology, and requirements. Rather than providing context and instructions with every interaction, custom GPTs encode your recurring patterns into permanent configurations you can invoke instantly.

Consider contexts where you repeatedly provide similar background information. If you frequently analyze market data using a consistent framework, create a custom GPT that knows your framework, understands your analytical priorities, and formats output according to preferences. If you regularly draft certain document types —proposals, reports, strategic memos—create custom GPTs that understand your organization's formatting requirements, tone preferences, and structural conventions.

Build custom GPTs through iterative refinement rather than attempting comprehensive configuration upfront. Start with the basic version handling core tasks. Use it for several weeks. Document what works well and where you're still providing repeated instructions. Update GPT configuration to incorporate those repeated elements. Progressive refinement creates truly useful custom tools rather than over-specified configurations that don't match real usage.

Most professionals find 3-5 custom GPTs cover 80% of recurring AI needs. More than that creates its own management overhead. Focus on high-frequency, high-value use cases where custom configuration provides clear advantage over generic AI interactions.

Workflow automation chains link multiple tools together so one action triggers a sequence of automated steps. This moves beyond isolated task automation to systematic process automation where human involvement focuses on decision points rather than execution steps.

Identify workflows involving multiple sequential steps where intermediate steps are largely mechanical. Client inquiry arrives via email, requires scheduling consultation calls, needs background research on their company, generates follow-up with relevant case studies. Each step is straightforward. The sequence consumes 45-60 minutes. Chaining these steps—automated email parsing, triggering calendar invitation sending, prompting AI research synthesis, generating personalized follow-up drafts—reduces total time to 5-10 minutes of human review and refinement.

Common automation chains for knowledge workers: content publication workflows where draft creation triggers editing checks, formatting application, platform-specific optimization, and scheduled posting. Project status workflows where task completion triggers progress report generation, stakeholder notification, and next-step planning. Analysis workflows where data collection triggers processing, visualization, insight generation, and distribution to stakeholders.

Build automation chains gradually. Start by automating connections between two steps. Verify reliability. Add the next connection. Progressive approach prevents creating complex systems that break unexpectedly with multiple failure points.

Document your automation chains including what triggers them, what steps they execute, where human review is required, and what constitutes successful completion. This documentation prevents confusion when chains produce unexpected results and enables troubleshooting when something breaks.

The Centaur Workflow achieves fluid movement between human and AI execution within individual tasks. The centaur concept emerged from Advanced Chess tournaments in early 2000s, where human-computer teams consistently outperformed both grandmasters alone and chess engines alone. Garry Kasparov, who helped pioneer this format after his historic matches against Deep Blue, observed that weak human plus machine plus superior process beat strong computer plus inferior process.

Apply this to knowledge work by identifying within complex tasks where human judgment adds most value and where computational processing provides greatest advantage. Design workflows that alternate between human and AI involvement seamlessly rather than treating tasks as entirely human or entirely automated.

For strategic analysis: AI processes large datasets to identify patterns and trends. Human reviews patterns to determine which merit deeper investigation based on business context AI lacks. AI conducts detailed analysis on selected patterns. Human interprets findings, considers implications AI can't assess, makes recommendations. AI drafts communication of recommendations. Human refines based on stakeholder knowledge and political considerations.

For complex writing projects: Human outlines structure and key points based on strategic objectives. AI develops the initial draft following outline. Human refines argument logic, strengthens key sections, adds examples requiring specific expertise. AI polishes language, improves flow, checks consistency. Human conducts final review focusing on accuracy, appropriateness, and strategic alignment.

The critical skill in centaur workflows is knowing when to hand off to AI and when to take control yourself. This judgment improves through practice and through explicit attention to where each transition point creates value versus friction.

That judgment—knowing when to hand off, when to take control—doesn't develop in isolation. It emerges through sustained practice across months and years. What separates professionals who maintain AI mastery from those who burn out chasing every update? Sustainable adaptation rhythms that fit their actual work context.

Four professionals demonstrate what this looks like in practice. They come from different industries, different temperaments, and different starting points with AI. What they share isn't technical aptitude or enthusiasm. It's discovering their own sustainable rhythm—the cadence of learning, experimentation, and refinement that maintains momentum without consuming their professional energy.

The Independent Consultant

Kat runs an HR consulting practice. Three years ago, she was one person doing the work of one person. Now she operates as "a firm of one that delivers like a firm of ten."

The key wasn't adopting every new AI tool that launched—it was developing problem-driven discovery that kept integration current without consuming client work time.

Her system: When she hits a bottleneck—a client request she can't handle efficiently, analysis that would take days, customization that's too time-intensive—that's when she investigates new capabilities. Not before. Quarterly, she blocks half a day for recalibration, reviewing which AI workflows still serve her and which need adjustment based on actual performance, not theoretical promise.

The result: She's increased client capacity by 400% without hiring. But more importantly, she doesn't feel like she's constantly chasing updates. The problem-driven rhythm means she can focus on organizational strategy—her actual genius—while AI integration handles volume work. She learns what she needs, when she needs it, because real bottlenecks force clarity about what matters.

KAT'S solo practice demanded individual discipline. Robert faced a different challenge: building sustainable AI practice across an entire team.

The Corporate Leader

Robert manages a fifteen-person engineering team at a manufacturing firm. His challenge wasn't personal AI integration—it was shifting an entire team's culture toward sustainable AI practice without disrupting production timelines or triggering resistance from veterans who'd built careers on technical expertise.

His breakthrough came from treating continuous adaptation as collective problem-solving rather than individual learning burden. He established monthly "workflow audits" where the team examines one major process together—not to evaluate performance, but to identify friction points. When someone surfaces a bottleneck, the team investigates whether new AI capabilities solve that specific problem. This isn't about staying current with AI trends. It's about solving real problems with whatever tools prove effective.

The discovery happens collectively. When one engineer hits a design visualization challenge that's slowing client approvals, the team investigates rendering tools together. When another struggles with materials specification documentation, they explore whether AI can automate the tedious cross-referencing. The learning load distributes naturally because it's tied to actual work obstacles, not abstract skill development.

The result: His team processes 60% more design requests than two years ago with the same headcount, and quality scores improved. But the real transformation is cultural. Team members stopped feeling threatened by AI and started feeling empowered by it—because they're solving their own problems, not implementing someone else's best practices.

Where Robert focused on collective problem-solving, Morgan's transformation was deeply personal—a creative professional confronting what felt like an existential threat to identity.

The Creative

Morgan is a residential architect specializing in custom homes. The initial resistance to AI was fierce—it felt like a direct threat to the creative work that defined professional identity.

What changed Morgan's perspective wasn't learning to use AI faster or more comprehensively. It was realizing continuous adaptation meant continuous creative expansion through direct experimentation.

Morgan now treats each major AI capability shift as a new medium to explore—similar to how architects learned CAD after starting with hand drafting, or BIM software after mastering 2D plans. When a new AI tool emerges that might expand creative possibility, Morgan experiments with it directly. Not reading about it or watching tutorials first—actually using it on low-stakes personal projects to understand its creative potential and limitations.

Quarterly recalibration specifically asks: "What can I design now that I couldn't design three months ago?" The question isn't about efficiency or productivity. It's about creative territory. Can I explore more radical structural concepts because AI handles engineering validation faster? Can I offer clients more design variations because iteration is no longer prohibitively time-intensive? Can I visualize custom details at a level of specificity that was previously impossible before construction?

The result: Creative output has become more ambitious, not less distinctive. Morgan takes on projects that would have been declined before—experimental structural concepts that require extensive iteration, rapid visualization cycles with clients exploring spatial options, personalized renderings at scale that help clients understand custom details before breaking ground. The creative voice didn't diminish. The canvas expanded.

MORGAN'S JOURNEY was emotional and experimental. Alan's conversion followed a more skeptical path, driven not by creative possibility but by competitive necessity.

The Skeptic

Alan is a senior partner at a law firm who initially dismissed AI integration as "fine for document review, irrelevant for real legal work." His conversion came gradually, driven by competitive pressure when younger partners at rival firms started winning clients partially through AI-enhanced efficiency.

His approach to sustainable practice reflects his systematic legal mind. He maintains what he calls "evidence-based adoption"—only integrating capabilities where he can verify quality improvement or time savings in his actual practice. He tracks outcomes obsessively: time-to-completion for standard documents, research accuracy, client satisfaction feedback.

His discovery method is rigorous testing. When he identifies a capability gap—research taking too long, contract drafting consuming too many billable hours, case law synthesis requiring excessive manual review—he investigates whether AI can address that specific gap. Then he tests. Small sample first. Verify quality against his own work. Measure time savings. Document results. Only after evidence confirms improvement does he integrate the capability into standard practice.

His quarterly recalibration follows formal protocol: review metrics, identify capability gaps, investigate specific solutions, test rigorously before committing. It's systematic enough to feel comfortable for his legal training while flexible enough to keep pace with relevant developments.

The result: He's recovered roughly 15 hours weekly from research and drafting work, redirected into client relationship building and strategic case development—work that actually differentiates his practice. More importantly for a skeptic, he has documented proof that his adoption decisions improve outcomes rather than just following trends. His evidence-based approach means he trusts his own integration choices—because he's verified them himself.

WHAT THESE FOUR share isn't enthusiasm or technical aptitude. It's the development of sustainable rhythms. Problem-driven discovery. Collective problem-solving. Creative experimentation. Evidence-based protocols. Each found their own version of continuous adaptation fitting their context, personality, and professional demands.

FUTURE-PROOFING FOCUSES development on capabilities where humans maintain durable advantage even as AI improves. You can't predict exactly which tasks AI will handle better in five years. You can identify skill categories where human value is highly resilient to technological advancement.

Complex interpersonal negotiation remains durably human. Negotiations involve reading subtle emotional cues, understanding unstated interests, building trust through authenticity, and making real-time tactical adjustments based on relationship dynamics. AI might eventually assist with some elements, but core negotiation itself requires human presence.

Physical world interaction and coordination stays human. Any work requiring direct physical presence, hands-on problem-solving, or coordination of physical activities maintains human advantage. Even with advancing robotics, human capacity for flexible physical problem-solving in unpredictable environments remains superior.

High-empathy relational work resists automation. Contexts where someone needs to feel genuinely understood—therapeutic relationships, mentoring, crisis support, transformative teaching—require authentic human connection. AI might support these activities but can't replace the human element.

Moral and ethical leadership requires human judgment. Decisions involving competing values, ethical dilemmas, organizational culture shaping, and moral courage draw on distinctly human capacities for meaning-making, value integration, and moral reasoning.

Invest professional development time disproportionately in these durable skill areas. As AI handles more routine cognitive work, your strategic advantage increasingly comes from capabilities AI can't replicate rather than from doing routine work better than AI.

10

LEADING YOUR RENAISSANCE

Every professional skill you develop has two potential returns on investment. The first return is direct: you use the skill yourself to produce better work, earn more, or save time. The second return is indirect but often more valuable: you teach the skill to others, which multiplies its impact, builds your reputation, and creates opportunities you couldn't access through direct application alone.

Most people focus exclusively on the first return. They develop skills, apply them personally, and measure success by their individual output. This approach works until you hit a capacity ceiling. There are only so many hours in your day, only so much work you can personally execute, only so much value you can create through direct application. Those who break through this ceiling recognize the leverage in the second return. When you help ten people develop a capability you've mastered, you've essentially created ten versions of yourself working in parallel. When you establish yourself as the person who understands something valuable, you get invited to conversations, opportunities, and decisions that weren't previously accessible.

. . .

LET'S address what may be running through your mind: Who am I to guide others? I've only been doing this for a few months. The real experts will see right through me.

These concerns reflect how we typically think about expertise. Traditional credentialing takes years: medical school, law school, apprenticeships, certification programs. We're trained to believe that teaching requires extensive formal preparation, and that offering guidance before achieving mastery is presumptuous at best, harmful at worst.

But AI integration isn't traditional expertise. The field changes so rapidly that someone who learned the fundamentals six months ago may be more useful to a beginner than someone who studied AI theoretically five years ago. Recent practical experience in a volatile field carries different value than accumulated expertise in a stable one.

Consider what you actually need to help someone starting their AI integration journey. They don't need comprehensive technical knowledge of neural network architecture. They don't need academic credentials in machine learning. They need to know: What does the first week actually look like? Which approaches work in practice? What mistakes should they avoid? What does "good" look like for basic applications?

You know all of this. You lived it. Your recency is an asset, not a liability—you remember the challenges clearly because you faced them recently, not through the haze of years of subsequent experience.

The qualification you need isn't "I know everything." It's "I know more than someone who hasn't started yet, and I can help them avoid the confusion I experienced." That's a lower bar than you're imagining—and a more useful one for the people you'd help.

Your framing: you're not an expert. You're someone who figured out a few things and is willing to share them. Nobody can argue with that. Nobody expects more than that. And that level of sharing provides enormous value to people who haven't figured those things out yet.

The fear of being publicly exposed as a fraud—that someone with genuine technical credentials will respond with devastating precision, revealing the gaps in your understanding—here's what actually happens when technically sophisticated people encounter practitioners sharing their experiences: usually, nothing. Technical experts occupy different territory than practical guides. They're interested in algorithmic improvements, model architecture, and research frontiers. They're not lurking in professional forums waiting to humiliate practitioners who share workflow tips.

When technical experts do engage with practical content, it's usually to add nuance rather than attack. A machine learning engineer might comment "That's a great practical approach—here's some context on why it works..." They're not threatened by practitioners any more than surgeons are threatened by physical therapists sharing recovery exercises. Different domains, different expertise, complementary rather than competitive.

The real risk isn't public humiliation. It's the quiet cost of staying silent. Every month you don't share your practical knowledge, those who would benefit from it struggle unnecessarily. They make mistakes you could have helped them avoid. They waste time you could have saved them. Your fear of theoretical criticism creates actual harm to real people who need what you know.

Focus on who you're serving, not who might theoretically critique you. The practitioners trying to figure out AI integration don't need protection from your imperfect knowledge. They need access to your practical experience. Serve them. Let the imaginary critics take care of themselves.

THE THOUGHT LEADER OPPORTUNITY

Authority in any field follows a predictable pattern: those who share their learning journey attract attention, build credibility, and eventually get invited to contribute at higher levels. This has always been true. What's different about AI integration is how few people are sharing genuinely practical guidance—and how many professionals desperately want exactly that.

The single most common mistake professionals make regarding thought leadership isn't sharing too early—it's waiting too long. Waiting for sufficient expertise means waiting indefinitely. In a field evolving as rapidly as AI, "sufficient expertise" is a receding horizon. By the time you feel ready by any reasonable standard, you've missed years of potential contribution. Worse, others who felt equally unready but shared anyway have established the positions you could have occupied.

The expertise-waiting mistake stems from comparing yourself to the wrong benchmark. You look at recognized AI thought leaders and measure your knowledge against theirs. The gap seems vast. You conclude you're not ready. But that's the wrong comparison. The relevant comparison is between your knowledge and the knowledge of people who would benefit from your guidance. Those people aren't looking for world-class expertise. They're looking for someone a few steps ahead who can help them avoid the confusion you've already navigated. Against that benchmark, you're ready now.

The cost of the waiting mistake is measured in help not provided—colleagues who struggled unnecessarily, organizations that made preventable errors, professionals whose anxiety about AI persisted because no one addressed it with practical experience they could relate to. Share before you feel ready. The feeling of readiness arrives through sharing, not before it.

Sharing your learning journey builds authority precisely because you're not claiming mastery—you're demonstrating progress. The thought leader positioning that resonates most strongly right now isn't the "AI guru" who has all the answers. It's the thoughtful professional who's one step ahead, sharing what worked, what didn't, and what they're trying next.

Here's a counterintuitive truth about building thought leadership: the less you claim authority, the more credibility you gain. Professionals who position themselves as fellow travelers—still learning, still experimenting, still questioning—attract more trust than those who project complete mastery. This creates the Permission Paradox. By explicitly giving yourself permission to share imperfect knowledge, you paradoxically become more qualified to share it. The acknowledgment of ongoing learning signals intellectual honesty that sophisticated audiences find reassuring. It says: "This person won't mislead me into overconfidence because they're honest about their own uncertainty."

When you share AI integration approaches, lead with your learning process rather than your conclusions. "I've been experimenting with X approach for the past three months and here's what I've found so far" lands better than "Here's the definitive approach to X." The first positions you as a credible guide; the second positions you as a target for anyone with different experience.

Warren Buffett and Bill Gates both follow what researchers have identified as the "5-Hour Rule"—dedicating at least five hours weekly to deliberate learning and skill development. Buffett reportedly reads 5-6 hours daily, explaining that "knowledge builds up like compound interest." This same principle applies to learning in public about AI integration. When you share your experiments, failures, and successes, you're building compounding authority with every post, conversation, and presentation.

Start documenting your AI integration journey in formats that match your communication preferences and professional context. For most professionals, this means some combination of three approaches.

Internal sharing through email updates, team meetings, or company Slack channels reaches colleagues who face similar challenges in familiar context. A monthly "AI Experiments Update" email to your team or department takes minimal time but positions you as the person tracking what works. Include: one experiment you tried, results you observed, what you learned, and what you're testing next. This four-part structure takes 15 minutes to write and establishes consistent thought leadership within your organization.

Professional community sharing extends reach beyond your immediate colleagues. LinkedIn posts documenting specific AI wins generate engagement because they address widespread curiosity with concrete examples rather than abstract claims. A post explaining "I used AI to cut my quarterly report preparation from 6 hours to 90 minutes—here's the exact approach" provides value that builds following. Focus on specificity: what task, what tool, what prompts worked, what adjustments you made, what the output quality looked like.

Peer-to-peer conversations with colleagues in similar roles create direct influence relationships. When someone mentions struggling with AI adoption, offer to share what's worked for you over coffee. These conversations build your reputation as the approachable resource—the person who actually knows how to make this work practically. Each conversation strengthens your network while refining how you explain your approach.

The "AI-Enabled" professional brand positions you as a forward-thinking leader without requiring technical expertise. Your positioning isn't "AI expert"—that would require credentials you may not have and compete with actual technologists. Your positioning is "strategic professional who has figured out how to use AI effectively." That's a category with relatively few credible voices and enormous demand.

Build this positioning through consistent signals in your professional presence. Your LinkedIn headline might include "helping [role/industry] professionals integrate AI strategically" or similar framing. Your speaking topics could address AI integration challenges your peers face. Your meeting contributions increasingly include references to AI-enhanced approaches you've implemented successfully.

Be specific about your area of AI application rather than claiming general AI expertise. "AI integration for strategic planning" is more credible than "AI expert." "Using AI to enhance client deliverables" is more defensible than "AI transformation specialist." Specificity builds trust because it sounds like experience rather than aspiration.

Teaching others accelerates your own mastery through the encoding benefit—and this isn't motivational rhetoric, it's cognitive science. When you prepare to explain something to others, your brain processes that information differently than when you're simply reviewing it for yourself. This phenomenon, which researchers call the "protégé effect" or "learning-by-teaching effect," has been replicated across educational contexts for decades. A study published in Applied Cognitive Psychology found that students who prepared to teach material significantly outperformed those who studied the same material for their own test. The preparation process—anticipating questions, organizing concepts for clarity, identifying potential confusion points—forces cognitive engagement that passive review doesn't achieve.

This means your hesitation to teach until you've "fully mastered" something has causality backwards. Teaching is one of the most effective methods for achieving mastery. When you prepare to teach something, you identify gaps in your own comprehension. When you answer questions, you develop new ways of thinking about familiar material. Each time you explain your AI integration approach to someone, you discover nuances in your own understanding. Each question you field exposes gaps you didn't know existed. The teaching itself produces the learning you thought you needed before teaching.

Identify teaching opportunities that match your current competence level and time availability. For minimal time investment: offer to lead a 20-minute "AI tools I'm using" segment in an existing team meeting. For moderate investment: organize a monthly lunch-and-learn where colleagues share AI experiments. For significant investment: develop a workshop for your department or professional association.

Structure teaching around problems and solutions rather than tools and features. "How to cut email time in half" engages attention better than "How to use ChatGPT for email." Professionals care about outcomes, not tools. Lead with the outcome, introduce the tool as the method.

Organizational influence follows demonstrated results and consistent contribution. When AI-related decisions come up at leadership levels, who gets consulted? Typically the person who has been visibly working on AI integration, sharing results, and building credibility through practical experience.

Here's a question that often unlocks thought leadership hesitation: If you could travel back nine months and give your past self guidance on AI integration, what would you say? You likely have immediate answers. You'd tell yourself not to worry about learning every tool. You'd recommend starting with specific problems rather than theoretical understanding. You'd share the three approaches that actually worked and warn about the two that wasted time. You'd address the specific fears that seem overblown now but felt paralyzing then.

That guidance you'd give your past self? Thousands of professionals are exactly where you were nine months ago. They have the same fears, the same confusion, the same paralysis. They need exactly the perspective you've gained.

The question reframes thought leadership from "What am I qualified to teach?" to "What have I learned that others haven't learned yet?" That's a lower bar—and a more useful frame. You're not competing with AI experts. You're helping people who were recently where you were. And for that role, your recent experience makes you more qualified than distant expertise ever could.

Position yourself for organizational influence by ensuring decision-makers know about your AI integration work. Send occasional brief updates to your manager highlighting AI-enabled wins. Mention relevant experiments in appropriate meetings. Offer to present findings to leadership if interest exists.

When organizational AI decisions arise—tool selection, policy development, team training—you want to be the person leaders think of as having a practical perspective. This doesn't require aggressive self-promotion. It requires consistent visibility combined with genuine expertise from your own integration experience.

LEADING TEAMS THROUGH CHANGE

Individual transformation is one thing. Helping a team transform is considerably more complex. Every person brings their own fears, resistance patterns, and learning styles. The psychological journey you completed over months must now somehow accommodate multiple people at different starting points moving at different speeds.

Psychological safety comes first—before any tool training, workflow changes, or efficiency discussions. This isn't just good leadership practice—it's evidence-based business strategy. Google's extensive research on team effectiveness, known as Project Aristotle, found that psychological safety was the single most important factor distinguishing high-performing teams from others. Without it, team members hide mistakes rather than learning from them, avoid asking questions that might reveal ignorance, and resist changes that feel threatening. With it, the same people experiment freely, share failures openly, and adapt to new approaches rapidly.

Your team members likely experience the same fears you did: identity erosion, skill obsolescence, imposter syndrome amplification. If you skip directly to "here's how we're going to use AI," you'll encounter resistance that has nothing to do with the technology and everything to do with unaddressed anxiety.

Open conversations about AI integration with explicit acknowledgment of concerns. Name the fears directly: "Some people worry AI means their expertise doesn't matter anymore. We're using AI to remove drudgery, not jobs. Your judgment, relationships, and strategic thinking become more valuable when routine work isn't consuming your energy."

Create space for team members to voice concerns without judgment. Ask specifically: "What worries you most about integrating AI into our work?" Listen without rushing to reassure. The goal isn't to eliminate concerns immediately—it's to demonstrate that concerns are welcome, heard, and taken seriously.

Share your own journey authentically, including initial fears, failed experiments, and gradual confidence building. When team members see that you experienced similar resistance and worked through it, they have a model for their own path. Your vulnerability creates permission for theirs.

Team protocols establish shared standards that remove individual decision burden and create consistent quality. Without protocols, every team member makes their own choices about when to use AI, how much to rely on it, and what requires disclosure—creating inconsistency and confusion.

Develop protocols collaboratively rather than imposing them top-down. Present draft guidelines, then facilitate discussion about what feels right for your team's context. Collaborative development increases buy-in, and it surfaces considerations you might miss.

Core protocol categories for most teams:

Disclosure standards specify when AI involvement requires mention. Typical approach: internal drafts and brainstorming don't require disclosure; client-facing final deliverables do require disclosure of AI assistance. Your team's standards might differ based on industry norms, client expectations, and organizational policy.

Quality verification requirements establish who reviews AI-generated content before use. The "human in the loop" principle from earlier chapters now becomes team policy. Define what review looks like: quick scan for obvious errors, or detailed verification of all facts and figures? Different output types might require different review intensity.

Data handling rules specify what information never enters AI systems. Client confidential data, personnel information, proprietary strategies, and other sensitive content should have explicit prohibition. Create specific examples: "Don't paste client financial data into ChatGPT" is clearer than "be careful with confidential information."

Role-appropriate applications clarify which tasks are appropriate for AI assistance at each role level. Junior team members might use AI for research and first drafts with mandatory review. Senior team members might have broader discretion. Make these distinctions explicit rather than assuming everyone understands the same boundaries.

Document protocols in a shared reference that team members can consult when uncertain. A one-page protocol summary prevents repeated questions and ensures consistent application. Review and update protocols quarterly as team experience and AI capabilities evolve.

Upskilling creates competence that reduces fear and enables contribution. Team members who don't know how to use AI effectively can't participate in AI-enhanced workflows. Their confusion becomes resistance. Their exclusion becomes resentment.

Structure upskilling around progressive skill building rather than comprehensive training. Start with one high-value use case that most team members can apply immediately. Once that's established, add the next use case. This progressive approach prevents overwhelm while building genuine competence.

The minimum viable AI skill set for most knowledge work teams includes: basic prompting for text generation (drafting, editing, summarizing), research assistance (finding information, synthesizing sources), and analysis support (data interpretation, pattern identification). Ensure every team member can execute these three categories competently before advancing to specialized applications.

Create practice opportunities with psychological safety. Designate specific time for AI experimentation where failures are expected and learning is the goal. A weekly "AI lab hour" where team members try new applications and share results normalizes experimentation without performance pressure.

Pair less confident team members with more experienced ones for peer learning. Direct instruction from a colleague often feels less intimidating than formal training. The teaching also reinforces the more experienced person's own learning.

Managing the adoption divide requires different approaches for different resistance patterns. Every team includes people who move at different speeds—some eager to try everything immediately, others reluctant to change anything.

For laggards who resist adoption, identify the root cause before applying solutions. Technical intimidation responds to patient skill-building support. Identity threat responds to reassurance about continued value. Philosophical objection responds to honest conversation about concerns and boundaries.

Set minimum participation standards while respecting individual pace. Perhaps everyone must complete basic AI training and use AI for at least one defined task. Beyond that minimum, allow variation in adoption speed. Some people will integrate more extensively over time. Others will maintain minimal use. Both are acceptable as long as team function isn't impaired.

For over-enthusiasts who adopt recklessly, establish guardrails without dampening initiative. These team members often produce creative applications but may skip quality verification or ignore data sensitivity. Channel their enthusiasm into testing and protocol development while ensuring their outputs meet quality standards before broader use.

Address conflicts between adoption speeds directly. The laggard who criticizes the enthusiast for "relying too much on AI" and the enthusiast who dismisses the laggard as "stuck in the past" both need redirection. Frame the diversity of adoption speeds as team strength rather than conflict source—different perspectives catch different problems.

RESPONSIBLE INNOVATION

With capability comes responsibility. You can now accomplish things that weren't possible before your AI integration. Some of those things you should do. Some you shouldn't. The difference often isn't obvious, and no external authority will make the decision for you.

Scale amplifies both benefits and harms—what's harmless at small scale can become destructive at large scale. The automation capabilities you've developed can be pointed in many directions. Some directions create genuine value. Others create volume without value, or worse, volume that actively harms others.

You could use AI to generate massive quantities of personalized outreach that recipients experience as spam. You could automate content production that floods channels with material indistinguishable from everyone else's AI-generated content. You could scale manipulative communication techniques that work on individuals but damage trust when discovered.

The ethical framework for scale decisions comes down to a simple question: Would I be proud of this approach if it were transparent? If competitors, clients, and colleagues could see exactly how you're using AI capabilities, would you explain it confidently or hide it defensively?

Apply the "front page test" to scale decisions. If your AI-enabled approach appeared on the front page of your industry's trade publication, would it read as innovative efficiency or as manipulation? If you're uncertain, that uncertainty itself is information worth heeding.

Define your own "won't do" boundaries before opportunities arise. When you've already decided that certain applications are off-limits, you don't have to evaluate each opportunity fresh while under pressure. Common boundaries: won't automate communications that deceive recipients about their source, won't generate content that misrepresents its nature, won't use AI to scale approaches I wouldn't do manually.

Job preservation should be a priority when efficiency gains create a labor surplus. AI integration often produces significant time savings. The question becomes: what happens to that saved time, and more importantly, what happens to the people whose work has been automated?

You likely don't control organizational headcount decisions directly. But you influence what happens within your team and what recommendations you make to leadership. Those decisions have real impact on real people.

When AI frees up capacity on your team, advocate for redeployment over reduction. Time saved on routine work enables higher-value contributions. People whose tasks were automated can be upskilled for work that requires human judgment, creativity, and relationships—work that benefits both them and the organization.

Build this philosophy into how you present AI integration results. Instead of "we saved 40 hours per week through automation," position results as "we freed up 40 hours per week that we're now investing in client relationships and strategic planning." The framing matters. It shapes how leadership thinks about AI's role and whether efficiency gains lead to growth or cuts.

If headcount reduction seems inevitable despite your efforts, advocate for thoughtful transition. Sufficient notice, retraining support, and honest communication respect the humanity of people affected. You may not prevent the decision, but you can influence how it's implemented.

Human-centric technology advocacy positions AI tools as serving people rather than replacing them. When organizational discussions turn to AI, you're positioned to be the voice arguing for applications that enhance human capability rather than substitute for human judgment.

Identify AI applications in your organization that seem oriented toward replacement rather than amplification. Customer service chatbots that frustrate rather than serve. Automated systems that create barriers to human assistance. Algorithmic decisions that remove human judgment from consequential choices. These applications often create short-term cost savings that damage long-term relationships and trust.

Raise concerns constructively rather than oppositionally. "Have we measured customer satisfaction with the automated system compared to human service?" is more effective than "This chatbot is terrible." Questions that invite data often shift conversations more effectively than statements of opposition.

Advocate for human oversight on consequential decisions. Any AI application that significantly affects people's lives—employment decisions, credit approvals, service access—should include human review. This isn't just ethical positioning. It's practical risk management. AI systems make errors that human judgment catches. Removing human oversight removes that safety mechanism.

Champion applications that genuinely improve experience for both employees and customers. AI that handles routine inquiries so human staff can focus on complex problems. Automation that eliminates tedious data entry so people can do more interesting work. These applications demonstrate that AI integration can be positive for everyone involved, not just shareholders.

Societal impact extends beyond your immediate organization to the broader community. Your voice in professional networks, industry associations, and public discourse contributes to how AI develops and how society adapts to it.

Participate in conversations about AI governance and ethics within your professional sphere. When your industry association discusses AI policy, contribute perspectives from practical implementation. When colleagues share concerns about AI's direction, engage substantively rather than dismissively. These conversations shape norms that eventually become standards.

Be honest about AI's current limitations even while advocating for its benefits. The credibility you've built through practical experience positions you to provide a nuanced perspective that counters both unfounded hype and unreasonable fear. "AI does X remarkably well, struggles with Y, and can't do Z at all" is more useful than unqualified enthusiasm or blanket dismissal.

Consider how your AI integration choices model behavior for others. When you demonstrate thoughtful, ethical, effective AI use, you provide an example that others can follow. When you cut corners or ignore concerns, you contribute to norms that make harmful applications more likely. Your individual choices matter beyond their immediate effects.

You've transformed your own practice over the past weeks and chapters—released routine work, expanded creative capacity, partnered on analysis, protected relationships, elevated strategy. Now comes the multiplication effect: when you help ten colleagues develop the same capabilities, you've created a tenfold impact beyond what you could achieve alone. The journey from individual competence to leadership influence is shorter than you think—it starts with one generous response to someone who needs what you know, then another, then another. Start this week with the five-day plan: identify your three biggest wins, help someone publicly, make an internal offer, document one workflow, and establish one sustainable practice. The positioning you build through consistent sharing creates opportunities that individual competence alone never generates, while helping others navigate challenges you've already solved.

CONCLUSION

The original Renaissance transformed how Western civilization understood human potential. Leonardo da Vinci embodied the "Renaissance man" ideal—integrating art, science, engineering, and human study rather than specializing narrowly. He painted the Mona Lisa and designed flying machines. He studied anatomy and wrote music. His interdisciplinary approach represented a new model of human excellence that valued integration over specialization.

That model applies directly to your professional future. The AI-amplified professional isn't a narrow specialist or a generic generalist. They're an integrator—combining deep domain expertise with AI collaboration capabilities with distinctly human skills in judgment, relationships, and creativity. That integration represents the new standard of professional excellence.

Work redefinition shifts professional value from effort to impact. In the pre-AI professional world, effort served as a reasonable proxy for value. If someone worked 60 hours on a project, that probably indicated more contribution than someone who worked 10 hours. Time investment correlated with output quality and quantity.

That correlation is breaking. AI-amplified professionals produce in 10 hours what unamplified professionals produce in 60

—sometimes at higher quality. If we continue measuring professional value by effort, the AI-amplified professional appears to be "working less" while actually contributing more. This creates cognitive dissonance for both the professional (who may feel guilty about "easy" work) and colleagues who measure value traditionally.

The resolution is explicit redefinition. Professional value equals impact delivered, not hours invested. Quality of judgment demonstrated, not quantity of effort expended. Problems solved and opportunities captured, not tasks completed and emails sent.

Make this redefinition explicit in your own self-evaluation. Instead of "I worked hard this week," assess "I delivered X impact this week." Instead of counting hours, count outcomes. This reframing reduces the guilt that many AI-amplified professionals experience while accurately reflecting their actual contribution.

Advocate for this redefinition in organizational evaluation systems. When performance reviews measure hours and activity rather than outcomes and impact, they incentivize the wrong behaviors. Contribute to conversations about how your organization evaluates performance, pushing toward outcome-based assessment that recognizes AI-amplified contribution accurately.

The amplified professional becomes the standard against which competence is measured. Just as computer literacy became expected for knowledge workers over the past decades, AI collaboration capability is becoming expected now. The professional who can't effectively use AI will eventually face the same disadvantage as the professional in 2010 who couldn't use email or spreadsheets.

This standard-setting happens gradually, then suddenly. For now, AI-proficient professionals stand out as exceptional. Within a few years, AI proficiency will be assumed. Those without it won't seem innovative —they'll seem incompetent. The window for early-mover advantage is open now but won't remain open indefinitely.

Position your AI integration capability as fundamental competence rather than special skill. It's not a bonus qualification. It's core professional infrastructure. This positioning protects against future commoditization of AI skills while accurately reflecting how professional expectations are evolving.

Help others develop this competence before the standard-setting is complete. The guidance you provide now—through teaching, mentoring, and team leadership—helps others achieve proficiency while the transition is still in progress. Once AI proficiency is assumed, catching up becomes harder and less rewarded.

Problem scale expansion becomes possible when routine work no longer constrains capacity. With AI handling research, drafting, analysis, and coordination, you have capacity for problems you couldn't previously address. Not just doing existing work faster—tackling bigger problems entirely.

What challenges have you deferred because they seemed too large for your available time? Strategic initiatives requiring extensive research? Cross-functional improvements requiring coordination across groups? Innovative approaches requiring experimentation before commitment? These challenges become addressable when routine work isn't consuming your capacity.

Actively seek larger challenges rather than accepting that saved time gets reinvested in more routine work. Parkinson's Law warns that work expands to fill available time. Unless you deliberately direct reclaimed capacity toward higher-impact work, the routine will creep back in—perhaps disguised as "using AI for more tasks" without genuine strategic advancement.

Set explicit goals for how you'll use reclaimed capacity. Perhaps one day per week goes to strategic thinking that routine work previously crowded out. Perhaps you take on a cross-functional initiative you previously couldn't have staffed from your own bandwidth. Perhaps you invest in relationship building that generates long-term opportunities beyond immediate deliverables.

The future remains uncertain, but your position within it is strong. Advanced AI capabilities continue developing. What seems impossible today may be routine tomorrow. No one can predict exactly how AI will evolve over the next decade, let alone longer timeframes.

What you can control is your relationship to that uncertainty. You've built a framework for evaluating new capabilities against your strategic needs. You've developed the learning habits that keep you current without drowning in updates. You've established your position as someone who integrates AI thoughtfully rather than either avoiding it or adopting it recklessly.

That positioning serves you regardless of how AI develops. If capabilities advance dramatically, you have the foundation to adapt. If development slows or stalls, you've already captured significant value from current capabilities. If new challenges emerge—regulatory, ethical, technical—you're positioned to navigate them thoughtfully.

The Professional Renaissance Method has taken you from anxiety to confidence, from overwhelm to competence, from fear of replacement to assurance of amplification. You know what to release to AI and what to protect for yourself. You know how to expand your creative capacity and elevate your strategic thinking. You know how to maintain relationships that AI can't replicate and judgment that AI can't replace.

That knowledge transforms not just how you work, but who you are professionally. You've become an amplified professional—more capable, more focused, more valuable than you could be alone. And you've developed the ability to help others achieve the same transformation.

The Renaissance you've led in your own practice is just the beginning. The Renaissance you can now lead for others—your team, your organization, your industry—extends that transformation far beyond yourself. That's the legacy available to you: not just personal transformation, but contribution to how your entire professional community navigates this transition.

The tools will keep changing. The capabilities will keep expanding. But the fundamental question remains the same: What should AI do for me, and what should I do myself?

You now have the framework to answer that question—for yourself, for others, and for whatever the technology becomes next.

Made in United States
Troutdale, OR
11/03/2026

56852692R00110